Living on the Edge

Tourism has an essential role in terms of contributing to the financial sustainability of protected areas. In addition, through effective and efficient benefit-sharing, tourism can positively impact numerous stakeholders within and beyond the protected area.

Living on the Edge: Benefit-Sharing from Protected Area Tourism highlights the complexity of benefit-sharing, the importance of identifying all relevant stakeholders, the challenges of ensuring equity and sustainability, and the critical importance of good governance. The evolution of benefit-sharing mechanisms over time also emphasizes a continuing need to evolve and adapt to each unique situation as much evidence indicates that little has changed for those living on the edge. Although this book focuses on benefit-sharing from protected area tourism, it is essential to acknowledge that along with these benefits are costs associated with tourism, including possible increased local prices, loss of access to land, human–wildlife conflict, and other related costs. The contributing authors agree that benefit-sharing must include good governance, accountability, equity, transparency, a broad reach of stakeholder engagement, and a robust combination of tangible and intangible benefits – with recognition that benefit-sharing systems need to be adaptive and evolve, as needed, according to the relevant situation.

The chapters in this book were originally published as a special issue of the *Journal of Sustainable Tourism*.

Susan Snyman, PhD, is Director of Research at the African Leadership University's School of Wildlife Conservation, and Research Associate with the School of Tourism & Hospitality at the University of Johannesburg, South Africa. She is also Vice-chair of the International Union for Conservation of Nature (IUCN) World Commission on Protected Areas (WPCA) Tourism and Protected Areas Specialist Group. Her research is focused on sustainable, diversified wildlife economies, including specifically, nature-based tourism and impacts on communities.

Kelly S. Bricker, PhD, is Professor and Chair in the Department of Parks, Recreation, and Tourism at the University of Utah. She is the Vice-chair of the Global Sustainable Tourism Council (GSTC); she coordinates the Capacity Building Working Group of the International Union for Conservation of Nature (IUCN) World Commission on Protected Areas (WPCA) Tourism and Protected Areas Specialist Group, and is Chair of the Central Wasatch Commission Stakeholders Group, Utah. Her research is focused on nature-based tourism, impacts on communities, and effective strategies for sustainable protected area and destination management. With partners in OARS, her husband, and 12 communities, she founded the conservation programme, 'Rivers Fiji'.

Living on the Edge

Benefit-Sharing from Protected Area Tourism

Edited by

Susan Snyman and Kelly S. Bricker

Routledge
Taylor & Francis Group
LONDON AND NEW YORK

First published 2021
by Routledge
2 Park Square, Milton Park, Abingdon, Oxon OX14 4RN

and by Routledge
52 Vanderbilt Avenue, New York, NY 10017

Routledge is an imprint of the Taylor & Francis Group, an informa business

British Library Cataloguing in Publication Data
A catalogue record for this book is available from the British Library

ISBN: 978-0-367-70290-8 (hbk)
ISBN: 978-0-367-70292-2 (pbk)
ISBN: 978-1-003-14551-6 (ebk)

Typeset in Myriad Pro
by Newgen Publishing UK

Publisher's Note
The publisher accepts responsibility for any inconsistencies that may have arisen during the conversion of this book from journal articles to book chapters, namely the inclusion of journal terminology.

Contents

Citation Information

The chapters in this book were originally published in the *Journal of Sustainable Tourism*, volume 27, issue 6 (May 2019). When citing this material, please use the original page numbering for each article, as follows:

Chapter 1

Living on the edge: benefit-sharing from protected area tourism
Sue Snyman and Kelly S. Bricker
Journal of Sustainable Tourism, volume 27, issue 6 (May 2019), pp. 705–719

Chapter 2

Revenue sharing from tourism in terrestrial African protected areas
Anna Spenceley, Susan Snyman and Andrew Rylance
Journal of Sustainable Tourism, volume 27, issue 6 (May 2019), pp. 720–734

Chapter 3

Economic impacts of tourism in protected areas of Brazil
Thiago do Val Simardi Beraldo Souza, Brijesh Thapa, Camila Gonçalves de Oliveira Rodrigues and Denise Imori
Journal of Sustainable Tourism, volume 27, issue 6 (May 2019), pp. 735–749

Chapter 4

Khanyayo village and Mkhambathi Nature Reserve, South Africa: a pragmatic qualitative investigation into attitudes towards a protected area
Dorothy Queiros and Kevin Mearns
Journal of Sustainable Tourism, volume 27, issue 6 (May 2019), pp. 750–772

Chapter 5

Strengthening governance processes to improve benefit-sharing from tourism in protected areas by using stakeholder analysis
Jasper Heslinga, Peter Groote and Frank Vanclay
Journal of Sustainable Tourism, volume 27, issue 6 (May 2019), pp. 773–787

Chapter 6

African tourism industry employees: expenditure patterns and comparisons with other community members
Susan Snyman
Journal of Sustainable Tourism, volume 27, issue 6 (May 2019), pp. 788–804

Chapter 7

Tourism development and the empowerment of local communities: The case of Mitzpe Ramon, a peripheral town in the Israeli Negev Desert
Joshua Schmidt and Natan Uriely
Journal of Sustainable Tourism, volume 27, issue 6 (May 2019), pp. 805–825

Chapter 8

Community involvement and tourism revenue sharing as contributing factors to the UN Sustainable Development Goals in Jozani–Chwaka Bay National Park and Biosphere Reserve, Zanzibar
Florian Carius and Hubert Job
Journal of Sustainable Tourism, volume 27, issue 6 (May 2019), pp. 826–846

Notes on Contributors

Kelly S. Bricker is Professor and Director of Parks, Recreation, and Tourism at the University of Utah. After completing her PhD at the Pennsylvania State University, she specialized in sustainable tourism and protected area management. Kelly has research and teaching interests in ecotourism, sense of place, community development, natural resource management, and the impacts of tourism. She has authored and edited books on sustainability which highlight case studies in tourism meeting environmental and societal issues, including *Sustainable Tourism & the Millennium Development Goals: Effecting Positive Change*; and educational texts such as *Adventure Programing and Travel for the 21st Century* and *Demystifying Theories in Tourism*. Dr Bricker serves the boards of the Global Sustainable Tourism Council, and the Tourism and Protected Area Specialist Group of the IUCN. With partners in OARS, communities, and her husband, she developed an ecotourism operation called 'Rivers Fiji'. Dr Bricker has served the academy for 22 years.

Florian Carius is Head of the Communications and Research Department of the Lower Saxon Wadden Sea National Park and Biosphere Reserve Authority, Germany. He lectures on tourism policy and nature-based tourism and previously worked as a scientist and advisor in international nature conservation.

Peter Groote is Associate Professor in the Department of Cultural Geography at the University of Groningen, the Netherlands. Research interests include heritage studies, economic history, historical GIS, places of death, and roadside memorials.

Jasper Heslinga currently works as Senior Lecturer-Researcher and Programme Manager at the European Tourism Futures Institute (ETFI), which is based at NHL Stenden University, the Netherlands. Heslinga obtained his PhD from the Department of Cultural Geography at the University of Groningen, the Netherlands. His research interests include: tourism planning, nature-based tourism landscape management, multi-level governance, institutional context, stakeholder analysis, historical content analysis, and policy analysis.

Denise Imori has a PhD in economics from the University of Sao Paulo. He works in the Sao Paolo State Department of Logistics and Transport.

Hubert Job is Professor of Geography and holds the Chair for Geography and Regional Science at the University of Würzburg, Germany. He is an elected member of the German Academy for Spatial Research and Planning and is an appointed member of the German National Committee for the UNESCO programme on Man and the Biosphere.

Kevin Mearns is Professor in the Department of Environmental Sciences at the University of South Africa. He specializes in the research areas of sustainability, sustainable tourism, community-based tourism, and community conservation.

Dorothy Queiros is Senior Lecturer in the Department of Applied Management at the University of South Africa. She specializes in community-based tourism, sustainable tourism development, and pro-conservation attitudes and behaviour towards protected areas. She holds a PhD in environmental management.

Camila Gonçalves de Oliveira Rodrigues is Researcher and Professor in the Department of Business and Tourism and the programme of sustainable development practices at the Federal Rural University of Rio de Janeiro. She has a PhD in environmental management and policies.

Andrew Rylance, PhD, is the United Nations Development Programme (UNDP) Head of Environment in Papua New Guinea. He is also the UNDP Chief Technical Advisor (CTA) in Papua New Guinea, specializing in environmental finance and protected areas. He previously worked as the UNDP CTA on similar projects in Seychelles and Mozambique.

Joshua Schmidt is an ethnographer and social researcher at the Ramon branch of the Dead Sea and Arava Science Center under the joint auspices of the Israeli Ministry of Science, Technology and Space, and the Ben-Gurion University of the Negev. His multidisciplinary research interests include contemporary Israeli popular and peripheral culture, ethno-linguistics, the sociology of tourism, ethnography, cultural shift, heritage conservation and management, disaster management, and response and recovery.

Susan Snyman, PhD, is Director of Research at the African Leadership University's School of Wildlife Conservation, and Research Associate with the School of Tourism & Hospitality at the University of Johannesburg, South Africa. She is also Vice-chair of the IUCN WCPA Tourism and Protected Areas Specialist Group. Her research is focused on sustainable, diversified wildlife economies, including specifically, nature-based tourism and impacts on communities.

Thiago do Val Simardi Beraldo Souza is an environmental analyst at the Chico Mendes Institute for Biodiversity Conservation. This work is based on his dissertation for his PhD in interdisciplinary ecology at the University of Florida.

Anna Spenceley, PhD, is an independent consultant focusing on sustainable tourism. She is Chair of the IUCN's World Commission on Protected Areas (WCPA) Tourism and Protected Areas Specialist Group (TAPAS Group), Director of the Global Sustainable Tourism Council, and Independent Advisory Panel member of Travalyst. See www.anna.spenceley.co.uk.

Brijesh Thapa is Professor in the Department of Tourism, Recreation & Sport Management at the University of Florida. His research theme is within the nexus of tourism, conservation, and sustainability.

Natan Uriely is Professor in the Department of Hotel and Tourism Management, Guilford Glazer Faculty of Business and Management at the Ben-Gurion University of the Negev, Beer-Sheva, Israel. His research covers various issues in tourism studies, mainly from a sociological perspective.

Frank Vanclay is Professor in the Department of Cultural Geography at the University of Groningen, the Netherlands. Research interests include social impact assessment and management, social sustainability, arctic development, sense of place, human rights impact assessment, social aspects of climate change, technology assessment, and cultural geography (social/human geography).

Living on the edge: benefit-sharing from protected area tourism

Susan Snyman and Kelly S. Bricker

ABSTRACT
Tourism in protected areas not only plays an essential role in terms of contributing to the financial sustainability of protected areas but through effective and efficient benefit-sharing can positively impact numerous stakeholders within and beyond the protected area. This paper provides a brief analysis of the evolution of benefit-sharing from protected area tourism, discussing tangible and intangible benefits and highlighting that revenue-sharing is only one way of sharing benefits from tourism and protected areas. The paper highlights the complexity of benefit-sharing, the importance of identifying all relevant stakeholders, the challenges of ensuring equity and sustainability and the critical importance of good governance. The evolution of benefit-sharing mechanisms over time emphasises a continuing need to evolve and adapt to each unique situation, as much evidence indicates that little has changed for those living on the edge. Although this paper focuses on benefit-sharing from protected area tourism, it is essential to acknowledge that along with these benefits are costs associated with tourism, including possible increased local prices, loss of access to land, human-wildlife conflict, and other related costs. Recommendations for future research are included to encourage an ongoing evolution and improvement in benefit-sharing mechanisms.

Living on the edge: an introduction

> "Most of the richest ecosystems host human populations which have lived in harmony with nature for centuries. These populations are now among the poorest in the world and depend entirely on the natural resources that surround them. If we want to conserve our natural capital and have access to genetic resources, we must recognize the contribution of native peoples and developing countries in maintaining these resources and share with them the benefits arising from their exploitation." Biber-Klemm and Martinez (2006, p. 4)

Society lauds protected areas (PAs) as one way of conserving biodiversity, ecosystem services and supporting human well-being. They are an integral part of sustainable development strategies, supporting the achievement of the Sustainable Development Goals (SDGs). One aspect of human well-being relates to the sharing of benefits from these PAs. For the purposes of this introduction, we define benefit sharing as *"a commitment to channel some returns, whether monetary or non-monetary, back to the range of designated participants: affected communities, source communities or source nations, participants in clinical trials, genetic disease patient groups"* (P2P

(2019), http://research.iftf.net/node/721). Therefore, benefits include both tangible and intangible benefits, ranging from direct financial benefits from the PAs, the ability to use the natural resources in the PA, to the ecosystem services resulting from the conservation within the PA. Many of these benefits are difficult to measure, and often even more challenging to communicate to those living around the PA. Research has shown that even the communication and measurement of tangible, financial benefits from, for example, tourism within PAs, is on an *ad hoc* basis with many governments being unaware of the total contribution to their economies (Rylance, Snyman, & Spenceley, 2017). So how then do we ensure that people support conservation if they are not aware of the value of the PAs and the benefits that they are receiving?

Regardless of the debate as to whether increased population growth around PAs, due to the creation of employment from PA tourism, or due to the ecosystem services provided (Igoe et al., 2008; Wittemeyer et al., 2008), poverty levels around PAs are often some of the highest in the world. Therefore, despite the primary role of PAs to conserve biodiversity, protected areas and issues of poverty are interlinked, creating both a moral and ethical responsibility for some action (Scherl et al., 2004). So what is the responsibility of PAs in the provision of benefits to people living on the edge and what role do they play in addressing local poverty reduction? Given many local communities' reliance on natural resources for survival (Sanderson & Redford, 2003; Snyman, 2013) and restrictions often placed on them due to the establishment of PAs, it is argued that PAs do have a responsibility to provide benefits and to reduce surrounding poverty (Scherl et al., 2004; Woodhouse et al., 2018). For example, in South Africa, South African National Parks, a parastatal organization managing the majority of PAs in the country, has a People and Parks programme to ensure the engagement of people in, and the provision of benefits to them, from the PAs. This trend has experienced steady growth, with the realisation that without the support of communities, PAs will not survive. As the late Nelson Mandela said *'ultimately conservation is about people. If you don't have sustainable development around these (wildlife) parks, then people will have no interest in them, and the parks will not survive"* (African Impact Foundation, 2018, p. 1).

The Vth IUCN World Parks Congress adopted as a principle the following statement: *"protected area establishment and management should contribute to poverty reduction at the local level, and at the very minimum must not contribute to or exacerbate poverty"* (Recommendation 5.29, in Scherl et al., 2004, p. 40). Also, at the 2005 Vth IUCN World Parks Congress, the theme of 'benefits beyond boundaries' and international support for more participatory governance and the sharing of benefits highlighted the linkages between poverty and protected areas.

Many would argue that 14 years later we are still struggling to ensure that these benefits beyond boundaries are perceived to be and being received. If conservation agencies were able to acknowledge and recognise the role of local communities and Indigenous people in the management of PAs, Kothari (2008) predicts there would be a dramatic increase in public support for (1) conservation and expansion of various kinds of protected areas; and, (2) a reduction in the conflicts that plagued many existing protected areas. However, this is not the reality to date. With an ever-expanding human population and increasing competition for land, we see more conflicts over land, with accompanying negative attitudes towards conservation. Human–wildlife conflict is often the perpetrator and the opportunity costs of conservation, which results in increasing pressure on PAs to provide benefits to those living in and around them. This competition between different land uses likely to continue to increase, primarily as developing countries focus on industry and development to develop their economies and reduce poverty. The issue of benefit-sharing from protected areas, and especially from the tourism in these areas, is going to become even more critical (hence the United Nations declaring 2018 the Year of Sustainable Tourism for Development).

Over the past four decades, there has been a 10-fold increase in the number of PAs globally. This increase has primarily been due to signatories to the Convention on Biological Diversity (CBD) striving to achieve Aichi Target 11:

"By 2020, at least 17 percent of terrestrial and inland water area and 10 percent of coastal and marine areas, especially areas of particular importance for biodiversity and ecosystem services, are conserved through effectively and equitably managed, ecologically representative and well-connected systems of protected areas and other effective area-based conservation measures, and integrated into the wider landscape and seascape" (Convention on Biological Diversity, 2019).

According to the latest report by the United Nations Environment Programme-World Conservation Monitoring Centre (UNEP-WCMC), globally we are heading towards the achievement of Aichi Target 11 in terms of the stated percentages of land and marine areas conserved. However, how well have we done in achieving effective and equitable management? In terms of equity, many communities living around PAs would argue that not much has changed.

Although the number and size of PAs has been increasing, the funding that is available to manage them has remained generally stagnant (Emerton et al., 2006) and in many cases has declined, resulting in significant shortfalls and insufficient means to ensure effective achievement of conservation and livelihood goals (Bovarnick et al., 2010; Watson et al., 2014). PA financing shortfalls are most intense in developing countries. Yet, the current financial needs for PAs in developing countries are thought to range between USD 1.1 billion, to cover the core operations, and approximately USD 2.5 billion to cover the very basic range of actions necessary to ensure effective management (Eagles & Hillel, 2008; Emerton et al., 2006; Lindsey et al., 2018; Rylance et al., 2017). These funding shortfalls highlight a potential challenge for PAs in terms of benefit-sharing, especially with local communities. If they are unable to fund necessary operational expenses, how will they then share benefits, particularly financial benefits, with those living on the edge? So, what is the solution?

To ensure that PAs are effectively managed through appropriate levels of funding, innovative and creative mechanisms to support this funding are needed. New approaches such as impact investing are growing in number (see for example Conservation Capital: https://www.conservation-capital.com), the promotion of legacy funds is another. Protected area tourism too can, if well managed, contribute to this shortfall. It is, however, not a panacea, despite a recent focus to expand and grow tourism in all PAs, it is not relevant or suitable in all PAs, and should only provide one option in a suite of funding options.

Sustainable tourism and sustainable development

The World Bank Group (2017) states that sustainable tourism is a proven tool for development which benefits communities in destinations as it is highly labour intensive, it facilitates new infrastructure development, it can help to fund conservation and can contribute to international understanding and peace. Recent trends in government (especially in Africa and Latin America), is to focus on tourism as a critical development tool reinforces perceptions that tourism is a powerful vehicle for economic growth and development, while at the same time has the potential to support conservation. This idea stems in part from the view, the environment may be a threat and a consumption opportunity for the poor, yet is a luxury (recreational) suitable for the wealthy. Also, this line of thinking supports a venue for wealth transfer from the affluent who want to enjoy the environment, to the poor, who frequently suffer from it (Snyman, 2013). PA tourism can create this avenue. Heavy reliance on tourism as the only driver of sustainable development and wealth transfer is, however, risky and may raise unachievable expectations. The management of the expectations of what PA tourism can deliver is a key challenge. An over-reliance on tourism as a development tool and sustainable financing option for PAs has resulted in high expectations of what tourism delivers. It would be prudent to manage these expectations, especially in terms of benefit-sharing: for whom and how many can benefit, and by how benefits are realized?

Globally, PAs receive roughly eight billion visits per year, of which more than 80% are in Europe and North America, not in areas where there is the highest poverty (Balmford et al.,

2015). Their study indicated these visits generate much more than is currently spent on conserving PAs, with approximately USD 600 billion per year in direct in-country expenditure and USD 250 billion per year in consumer surplus (Balmford et al., 2015). Research has demonstrated an increase in nature-based tourism can significantly impact the operational management of PAs, with the related increase in economic benefits supporting the financial sustainability of conservation efforts (Kim et al., 2019). So how do we ensure that these growing receipts from tourism, are funding conservation, and shared equitably with stakeholders? This special issue provides evidence of ways in which benefits are being shared, the challenges of benefit-sharing and how to overcome these.

We continue our introduction of this special issue through a brief analysis of the evolution of benefit-sharing from PA tourism; identifying some of the issues and challenges, and utilizing themes identified by papers within this issue, including: stakeholder engagement; governance; and the impacts of benefit-sharing; and conclude with a brief analysis of the papers within this issue. Although we focus primarily on benefit-sharing from PA tourism, it is essential to acknowledge that along with these benefits there are often costs associated with tourism, including the potential for increased local prices, loss of access to land, human-wildlife conflict, and so on.

Evolution of benefit-sharing

Very early on in the evolution of ecotourism or nature-based tourism, PA managers, non-governmental organisations, communities, and other relevant stakeholders have recognised separating people from parks was not a sustainable strategy. Hard lessons were learned, with the realisation, in order to conserve ecosystems, people living in and around protected areas must receive benefits. Early benefit models highlighted the utilisation of local procurement, continued access to natural areas, local revenue-sharing, and local employment as typical benefit-sharing processes. Many have argued that these processes were far more complex and required substantial considerations for effective benefit-sharing to be realised. In essence, benefit-sharing must be ample enough to contribute to livelihood needs and compensate residents living next to a protected area (Adams et al., 2004). The primary mechanism was financial and primarily through revenue-sharing (Ahebwa et al., 2012; Archabald & Naughton-Treves, 2001; Umuziranenge & Muhurwa, 2017). This mechanism was meant to demonstrate a simultaneous connection with local communities, which in turn assisted in meeting conservation goals (Ahebwa, van der Duim, & Sandbrook, 2012). More recently, those embracing of successful revenue-sharing have included: long-term institutional support; appropriate identification of target communities and project types; transparency and accountability; adequate funding and the development of realistic expectations (Spenceley, 2014).

Increasingly, the complexities of tourism systems coupled with the complexities of PA management call for a socio-ecological system approach to understanding nuances of various types of benefits. New models and partnerships are evolving to address these complexities. To understand the evolution of benefit-sharing related to protected area tourism, we have organized our conception of this evolution through three prominent themes identified by papers within this special issue: stakeholder engagement, governance, and the impacts of benefit-sharing.

Stakeholder engagement

Historically, much of the focus of benefit-sharing from PA tourism has been financial, which does not constitute the full impact of this socio-ecological system. If stakeholders are made aware of all the benefits of PAs and the importance of PAs to human well-being, then the scope of benefit-sharing would increase, and more stakeholders would be benefitting. This notion is supported by the Protected Planet Report 2016 (UNEP-WCMC and IUCN, 2016), which states that ensuring a more sustainable future for people and the planet will require greater recognition of the vital role that PAs fulfill in underpinning sustainable development. Hence, there is also a need to strengthen the communication of the benefits of PAs across all sectors of society to help demonstrate the economic and social values of PAs to existing and future generations (Aichi Biodiversity Target 1) (Bruyere et al., 2009; UNEP-WCMC and IUCN, 2016).

Research has demonstrated that tourism can provide value to society through the potential economic, environmental, and social benefits it can deliver (Snyman & Spenceley, 2019; Mbaiwa & Stronza, 2010). An understanding of these benefits, both tangible and intangible, and how they are shared, is essential to illustrate their value. If PA tourism is, however, required to fund conservation in PAs, due to the shortfalls discussed above, then how much of the benefits remain to share with other stakeholders?

Over the years, research has shown that the perception of benefits can sometimes outweigh the actual receipt of benefits. Equity is an essential aspect in terms of the long-term sustainability of protected area tourism related to benefit-sharing. Franks, Booker, and Roe (2018) discuss equity in terms of conservation goals overall, but the concept equally applies to benefit-sharing from conservation and PA tourism. Franks et al. (2018) define equity as "fairness" or "social justice," which has three dimensions: recognition, procedure, and distribution. Equity is also related to governance, which is about power, relationships and accountability (Franks et al., 2018; Schreckenberg, Franks, Martin & Lang, 2016). A key aspect of governance is decision making which frequently involves tough choices between many competing objectives (trade-offs) (Franks et al., 2018). An understanding of local community dynamics, hierarchies and power-sharing are also essential to develop methods to ensure that the most marginalised and most impoverished in communities are also included in benefit-sharing mechanisms (Kothari, 2008). For stakeholders, consultation, joint decision-making, and advocacy are needed to move towards equity (Kothari, 2008).

An essential part of recognising the benefits of PAs and the role that they play is understanding the stakeholders involved and ensuring that the importance of PAs is communicated to all stakeholders. In this regard, the World Bank Group (2018) highlights the importance of carrying out a stakeholder assessment to understand the requirements of all major stakeholders in PA tourism. They also provide detailed information on the relevant stakeholders, their roles, goals, and challenges (World Bank Group, 2018, pp. 52–53). Heslinga, Groote, and Vanclay (2018) in this issue highlight the importance of stakeholder involvement in ensuring adaptation, flexibility, an understanding of the enabling conditions and constraints: all crucial factors in facilitating more successful benefit sharing programmes.

There are numerous diverse stakeholders engaged in tourism in PAs including: government, local communities, non-governmental organisations, and the private sector, amongst others. Many of these stakeholders have different, sometimes conflicting, objectives for being involved in PA tourism, and some only appear once the benefits start appearing. Local communities, in particular, frequently bear the costs of living adjacent to PAs through the negative impacts of human-wildlife conflict, as well as the opportunity costs of not being able to use the land for agriculture or other development (Snyman, 2013). It is essential, therefore, that local communities, those living on the edge, receive benefits from conservation in PAs and the associated tourism. These benefits can be through direct receipt, but recently there has been a focus on empowering communities to engage in tourism through joint ventures and other collaborative management models (Baghai et al., 2018). Schmidt and Uriely (2018) in this issue, however, argue that the notion of empowerment is a different and relative construct, with some aspects of empowerment being reinforced due to tourism development, other aspects deteriorating as a result of this dynamic. In order to ensure equitable power relations when communities do engage in tourism, however, there needs to be concomitant capacity building to enable them to understand the complexities of tourism as a business model and the associated potential risks, challenges and benefits.

Within the concept of stakeholder engagement, an assortment of influences empowers or disempower individuals and communities at large. Psychological empowerment, for example, enhances self-esteem, with the idea that access to economic gains (i.e., employment and cash) leads to an increase in status (Scheyvens, 1999). The creation of a sense of pride in a community (Mbaiwa & Stronza, 2010; Xue, Kerstetter, & Hunt, 2017), and increased access to the typically underserved such as youth and women (Roe & Urquhart, 2001) are evolving into benefits and being recognised by diverse stakeholders. Social empowerment enriches community cohesion among stakeholders, with economic returns supporting community development activities (Cole, 2006; Mbaiwa & Stronza, 2010; Scheyvens, 1999). And, political empowerment is described as "the community's political structure, which fairly represents the needs and interests of all community groups, provides a forum through which people can raise questions ... and have their concerns addressed" (Scheyvens, 1999, p. 247). These empowerment facilities are deemed critical within the broad spectrum of elements of successful benefit-sharing attributes.

Through diverse PA benefit-sharing situations, connections between the leadership within a community and residents are salient to successful benefit-sharing programs. For example, these connections may influence the level of political empowerment. For example, residents living adjacent to Kibale National Park in Uganda identified the inclusion of community stakeholders in the decisions of types of projects to assist in minimizing the impacts of wildlife on adjacent park communities, as necessary to their perception realizing benefits (MacKenzie, 2012). In Fiji, stakeholder engagement was responsible for how a private protected area managed tourism employment, using a mix of traditional and contemporary employment guidelines (Bricker, 2001). Globally, researchers are exploring ways in which tourism in and near protected areas are fostering cultural preservation, sustainable behaviour, and perceived control of tourism enterprises (Lackey & Bricker, 2019). These new ways of understanding various influences and outcomes may reveal innovative ways of conceptualizing localized benefits and elements of empowerment relevant to long-term PA tourism and conservation successes.

While the literature has established the importance of stakeholder engagement, there is also evidence that the mechanisms that enhance community agency in the development of tourism and empower stakeholders are less clear. Several barriers to establishing community agency can thwart levels of active stakeholder engagement. Bergstrom and Harrington (2018) identified the importance of stakeholders' control over change within a community and adjacent lands. This idea harkens back to the critical inclusion of access to training, building capacity including access to education, information, and continuous training, and ensuring empowerment of marginalized groups (Lackey & Bricker, 2019; Panta & Thapa, 2018; Ramos & Prideaux, 2014; Su, Wall, & Xu, 2016; Thondhlana, Shackleton, & Muchapondwa, 2011).

Beyond economic benefits (i.e., employment opportunities and direct revenue), successful benefit sharing programmes should encourage and address capacity building and skills training opportunities amongst various stakeholder groups (Archabald & Naughton-Treves, 2001; Scheyvens, 1999). These efforts are designed around the specific needs of the community and ideally lead to a level of empowerment. Scheyvens (1999) outlined a framework for empowerment relative to PA tourism within local communities. Signs of empowerment and disempowerment were identified beyond economics – to include social, psychological, and political factors. These factors influence levels of stakeholder engagement (social), self-esteem (psychological), and overall governance (political) factors. Lack of empowerment may impact the degree to which benefit-sharing can address localized and institutionalize challenges.

Governance

In considering good governance, making investments into communities within and around PAs could deliver more in terms of human well-being and poverty alleviation, and better conservation, if the investments are focused on enhancing equity rather than directly improving livelihoods or reducing poverty (Franks et al., 2018). Highlighting an important policy issue and perhaps a better solution is to combine enhancing equity in communities and promoting effective and equitable benefit-sharing. Further research focusing on equity and governance related to tourism and protected areas is vital to the long-term sustainability of conservation and related tourism. Future research should also focus on assessing different aspects of, and best practices for, good governance, and how these can be incorporated explicitly into benefit-sharing mechanisms to promote more significant equity.

Each community is unique with its nuances in terms of structure, roles, and responsibilities, and should be considered within benefit-sharing schemes. Besides, considerations embracing local cultures and traditions which are continually evolving may have a significant impact on roles within communities and the understanding of equity (Kothari, 2008). Policy and legislation need to be adaptive to these changes.

The governance of benefit sharing structures and programmes impact practical benefits and types of benefit-sharing participation. We are seeing increased advocacy for strong institutional support, with efforts by policymakers and associated government policies, processes and systems now called upon to lessen the barriers to community participation (Scheyvens, 2002). Therefore, it is not only about what governance is in place but the quality of governance which is also a key challenge related to benefit-sharing. Ensuring good governance can, therefore, play a key role in overcoming many of the challenges related to benefit-sharing (See Heslinga, Groote, and Vanclay in this issue who look at strengthening governance to improve benefit-sharing).

There is also evidence that institutional frameworks can promote or detract from ways in which benefits come to fruition. Public versus private protected areas can impact how agreements, partnerships, formal and informal structures are developed. For example, within the Republic of Fiji, a community-private partnership programme called Rivers Fiji, located within the rural highlands of Viti Levu has been operating within a private protected area now for nearly 20 years. The institutional structure impacts directly how local communities directly receive revenue from the project. As a privately protected river corridor, the Upper Navua Conservation Area is Fiji's first and only Ramsar Site (i.e., Wetland of International Importance), but also is protected through a lease for conservation (see Bricker, 2001). Despite the opposition of the local government leasing authority (Native Land and Trust Board), the lease for conservation was set up with not only a lease payment but revenue from each visitor to the canyon.

In addition to clearly identified employment sharing benefits, on-going revenue sharing is parallel to both the company's and partnerships success. With the advantage of hindsight, had Rivers Fiji set this up with the traditional one-time lease payment, as the tourism enterprise became increasingly successful, benefit sharing at the local community level would have remained stagnant. Instead, how the lease payment was set, local communities benefit continuously, therefore realizing the full value of this protected area partnership every time a tourist travels through the river corridor (see Bricker, 2001).

Increasingly evident is that well-structured partnership agreements and governance structures, with transparent benefit-sharing processes, provides opportunities for communities to connect to economic benefits and employment opportunities (Archabald & Naughton-Treves, 2001; Bricker, 2001; Mbaiwa & Stronza, 2010; Roe & Urquhart, 2001). While there are vast differences in these types of agreements, common to all must be a transparent structure of engagement and procedures for fair distribution of opportunities. In the case of Rivers Fiji, the land-owning group utilized traditional structures and practices for rotating the employment benefits between different groups, this has worked well, and there is complete buy-in by all landowning groups for this structure involved in the Upper Navua Conservation Area (Bricker, 2001, 2013). Also, these structures or processes have increased the types of stakeholders engaged in the various projects, increasing awareness and ultimately more widely spread opportunities for benefit-sharing within and across communities, including educational opportunities, healthcare, and tourism support services (i.e., drivers, food supply, and infrastructure development).

Research has also indicated an understanding of the system in which communities operate have significant impacts on effective benefit-sharing strategies. For example, in Kibale National Park (Uganda), residents garnered more benefit from being able to work on infrastructure to minimize damage from elephants and wildlife, than the revenue from park fees (MacKenzie, 2012). Hence recognizing the value of work to this specific community in establishing long-term perceived benefit was useful in determining types of benefit-sharing alternatives, when the fees alone were of lesser value (Winchenbach, Hanna, & Miller, 2019). Hence, understanding particular community needs are essential to understand benefit-sharing mechanisms or processes, and while economic benefits have been shown as critical, there are multiple ways in which benefits may be fulfilled.

Innovation in policy and governance is taking place through a range of new types of partnerships and engagement structures. For example, adaptive co-management of PAs may be a way to address the "intractable challenges facing protected areas and sustainable tourism such as conflict, complexity and uncertainty" (Plummer & Fennell, 2009, p. 163). With co-management structures, there may be increased opportunities for local empowerment and innovation, positive perceptions of engagement, perceived control, and sustainable behaviors. While specific research on co-management of tourism and PAs is sparse, there is some indication of positive benefits because of a level of participation by communities, including equitable benefit distribution and relatively high levels of communication (Mutanga, Muboko, & Gandiwa, 2017; Puhakka, Cottrell, & Siikamäki, 2014). Besides, when there is a clear plan for engagement and succession, there is continuity, identified as a positive attribute in benefit-sharing schemes (Bricker, 2013; Spenceley, 2014). However, indicators of poor relationships or even conflict between PA managers and communities have included low tourism involvement, the presence of human-wildlife conflicts, and mutually negative perceptions between local communities and PA managers (Bango & Xelelo, 2017; Lackey & Bricker, 2019).

Impacts of benefit sharing

The sustainability of protected area tourism in part depends on its ability to play a role in biodiversity conservation and, through this, assisting the PA in achieving its objectives and ensuring effective alignment with the values of the PA (Hockings et al., 2006). Umuziranenge and Muhurwa (2017) established that tourism revenue sharing and related benefits could serve as an incentive for conserving Nyungwe National Park in Rwanda. Tourism managed well can act as an incentive for conservation by stopping illegal or destructive activities. Makame and Boon (2008) have also suggested that in order to engage full benefits of natural resource tourism, precise mechanisms from institutions such as Forestry and Parks departments and others must be in place for benefit sharing to occur. Some research has revealed a perceived divide between communities and natural resource tourism management.

In a study of Purnululu National Park and Warmum Community in Australia, results disclosed a perceived separation of Indigenous people from the Park and tourism occurring in the region (Strickland-Munro & Moore, 2013). Besides, Indigenous people's cultural connections to their landscape contributed to their local stewardship and "custodial" expression of responsibility towards the Park, tourists, and nearby landscapes (Strickland-Munro & Moore, 2013). Pragmatically, this research revealed that Indigenous people identified several constraints in actually accessing the park and park tourism, due to lack of transportation, need for four-wheel drive, and driving capability (lack of license), as a reason for isolation from the Park. Ultimately, the Park's management team faced competing for demands "to ensure access for local Indigenous people, while retaining the "wilderness" experience for tourists and for which the park is renowned" (Strickland-Munro & Moore, 2013, p. 35). Not surprising, their research found not all community members wish to engage in tourism and that under-engagement may actually be a successful outcome of sustainable tourism—"individuals may choose not to be involved in economic opportunities available from Park tourism, preferring instead to prioritize non-financial, intrinsic benefits such as the maintenance of traditional practices" (p. 36). Understanding the desires of residents and stakeholders is an essential consideration in conceptualizing the sustainability of tourism and benefit-sharing within varying and diverse local contexts.

For example, tourism industries have the ability to share benefits with communities and other stakeholders through the provision of health services, improved water supply, increased educational support mechanisms, increased diversity of employment opportunities, infrastructure development, sustainable resource management, improved local attitudes towards natural resources conservation goals (Ahebwa et al., 2012; Bricker, 2013; Makame & Boon, 2008;Snyman, 2016). Further, somewhat related, others have argued for processes that increase awareness and interest by local communities on regional issues to encourage equity and empowerment (Roe & Urquhart, 2001; Scheyvens, 1999; Tosun & Timothy, 2003). As noted previously, research has shown some indication that benefits do not always have to be tangible to have a positive impact (Mbaiwa & Stronza, 2010; Scheyvens, 1999). For example, intangible benefits such as capacity building, skills development, and increased community-level decision-making have demonstrated significant social and human capital positive, long-term impacts.

In summary, research demonstrates that clear, equitable transparent benefit-sharing systems ensure a higher likelihood of success for communities and sustained tourism business success as well as meeting conservation goals. Benefit-sharing then becomes part of the give and take of a complex socio-ecological system, whereby strategies and implementation goals must be adapted to address ever-changing cultural (Mbaiwa & Stronza, 2010), social (Xue et al., 2017) and ecological (Bosak, 2016; Mbaiwa & Stronza, 2010) systems.

As a framework for understanding complex systems, researchers have suggested both tangible (e.g., revenue and economic returns, infrastructure, health and ecological) and intangible aspects of communities, such as empowerment (e.g., psychological, social, and political) are considered and influence the effectiveness of benefit sharing and tourism development. With such an emphasis on revenue generation, poverty alleviation, and various forms of capacity building (see United Nation's Sustainable Development Goals), benefit-sharing is a complex interaction which deserves increased attention. For example, consider the opportunity for significant innovation at the policy level, which in turn may have significant implications at the community level. Governments could consider tax incentives and support schemes for tourism enterprises that demonstrate increased benefit sharing structures and programmes. For those tourism enterprises that contribute to the protection of biodiversity, and build capacity in communities living on the edge of PAs, incentive programs (tax or otherwise) could be implemented.

Despite a focus on the sharing of benefits from PA tourism, it is essential also to consider numerous challenges related to benefit-sharing. Spenceley (2014) identified six key challenges with benefit-sharing from PAs. These include: (1) the value for money per person is small if divided among a large number of people; (2) benefits of social infrastructure (e.g., schools, water, infrastructure) are not always associated with the conservation or tourism, (3) those who benefit are not necessarily the same as those who experience the costs of conservation, for example, human-wildlife conflict, loss of access to land, (4) poorest residents are often not the beneficiaries, (5) community entities may not have the capacity to partner with other stakeholders or to agree on benefit-sharing processes, and, (6) legislation may constrain benefit-sharing processes(adapted from Spenceley, 2014). In a reinforcing cycle, stakeholder engagement is intrinsically linked to equity, and assessing equity involves assessing the quality of governance in terms of principles of good governance (Franks et al., 2018).

Stakeholder engagement is considered key to effective benefit-sharing. Inadequate communication of the benefits to stakeholders, and a lack of capacity to manage funds received from protected area tourism within community structures can highlight the adverse effects when governance is less than adequate (Makame & Boon, 2008; Spenceley, 2014). A primary challenge has been to identify who the stakeholders are that *should* benefit. This group can be large and diverse within the realm of tourism systems, with some stakeholders being easily identifiable, while others are less obvious. Temporal considerations are also at play, whereby stakeholders eventually emerge as tourism develops. Stakeholders may also have different and sometimes conflicting reasons for engaging in tourism, which adds to the complexity. For example, ensuring linkages are established between the benefits received and the associated tourism and protected area is yet another challenge (Snyman, 2013). Another challenge is determining how benefits are shared within and amongst stakeholder groups. For example, whether benefits will be individual or collective and who will determine this and manage the distribution (Bricker, 2001; Tumusiime & Vedeld, 2012). What is clear however is, benefit-sharing resulting from PA tourism is challenged by a lack of empowerment and participation of stakeholders, coupled with the realization that there will be costs and the associated benefit-sharing programs must be at a scale to mitigate these potential costs (Tumusiime & Vedeld, 2012).

Indeed, without this apparent link, there may be a little perceived value associated with tourism and conservation and therefore no inclination to protect or ensure its long-term success. To some extent, most of these challenges are in some way faced in the papers in this special issue and can serve to undermine the potential positive impacts of benefit-sharing from protected areas and tourism. These challenges are not new or unique to certain places or communities. It is crucial, therefore, to seek ways to overcome these challenges and ensure more effective and efficient benefit-sharing. Active institutional structures and support, the establishment of good governance practices and policies, clear benefit distribution plans, community engagement, empowerment, promoting more extensive value chains, increasing indirect and induced impacts, all play a role in addressing the challenges and promoting more positive, sustainable benefit-sharing.

Living on the edge: the special issue

This special issue includes six papers identifying diverse aspects of benefit sharing as well as scale and regional differences. Findings support benefit-sharing as a socio-ecological system which includes tangible and intangible concepts. The papers build upon previous research focused on revenue and economic impacts, and empowerment through agreements and structures. Also, the papers address various stakeholder engagement; the importance of unique contexts and level of community participation; and the support for a systems approach to understanding the complexities associated with benefit-sharing from PAs, utilizing a community capitals framework to address community context and empowerment comprehensively.

Economic benefits continue to be an essential focus of benefits, yet must have structures and functional systems in place to realize their full potential. Revenue sharing should be sufficient and regular, as well as timely, and offset costs experienced by communities such as crop damage, livestock depletion, and restrictions on resource use. Revenue sharing specifically from tourism enterprises should be mindful of implementing specific agreements, inclusive of set leasing fees, with accountability for growth annually (Rylance, Snyman, & Spenceley 2017; Snyman, 2018; Stone & Nyupane, 2014). Collectively, the results also suggest the importance of the context in which specific communities operate (Rylance, Snyman, & Spenceley, 2017; Snyman, 2018; Souza, Thapa, & Inmore, 2018). Without mechanisms for incentives and revenue generation, communities will lack buy-in and acknowledgment of the benefits associated with PA conservation initiatives (Souza et al., 2018, this issue). Most authors within this special issue agree that these relationships and situations that may or may not provide incentives for communities are context specific (Rylance, et al., 2017). The authors also found that overall revenue sharing is context specific, in that some communities experienced more significant losses (i.e., livestock loss) due to their proximity, with other communities experiencing a net gain. These results have direct implications for policy development that is adaptable and context specific. In tandem with active policy formation, authors also noted the increased need for strong technical and institutional support of policies and greater collaboration and understanding between park authorities and local communities. Concerning local communities, they discovered the absence of consensus, or perhaps unity in understanding within a community, affects the bargaining power and ultimately benefits for communities.

How agreements, structure, and governance are designed and initiated adds to the understanding of specific context, levels of empowerment, enabling strategies, and stakeholder engagement (Snyman, 2018; Heslinga, Groote, & Vanclay, 2018; Schmidt & Uriely, 2018). The use of the community capitals framework by Stone and Nyaupane (2014) is a step forward in addressing the complexities of systems, discarding the cause and effect relationships for a broader solution-oriented impact. Reinforcement of community cohesion, environmental awareness and education, and enjoyment are intangible benefits that can lead to a new sense of responsibility towards protected areas, and ultimately strengthening an understanding and support for conservation efforts (Queiros & Mearns, 2018).

Moving forward

The papers all highlight the need for comprehensive stakeholder engagement and frameworks to address the complexity of benefit-sharing. These ideas ensure stakeholders are aware of all benefits and the value of tourism and conservation in and around PAs, as well as the importance of efficient and effective governance structures to manage tourism and related benefit-sharing mechanisms. Collectively these papers provide a thorough overview of several different issues related to PA tourism and the associated benefit-sharing. They also highlight that there is still much to be understood and researched in terms of benefit-sharing, including more on the multiplier and induced impacts, understanding the role of equity and governance, and acknowledging factors, which impact both tangible and intangible benefit-sharing outcomes.

Adding to the need for increased understanding within each of these areas, many have identified the tourist as key stakeholder often left in the margins when examining its role in supporting PA tourism benefit-sharing. Where does the tourist fit in the context of fees for services, and entry fees, philanthropic activities, or perhaps voluntourism initiatives? Further, how will ecosystem services, inclusive of access to fresh water, food, quality of life, and other health and well-being factors, impact benefit-sharing in tourism and protected areas? Does PA tourism provide an increased opportunity as biodiversity and the health of ecosystems increases?

Authors in this special issue agree that benefit-sharing from PA tourism is complex, with many challenges to ensuring equity, accountability, and sustainability. Despite this, there are also numerous positive impacts from benefit-sharing which are highlighted in the papers, including employment, financial benefits, empowerment, infrastructure, and services. Found as critically important, are both tangible and intangible benefits, with intangible benefits often seen as having more extensive, longer-term impacts and frequently reaching more people. Illuminated within these papers, is there is not one benefit-sharing system that is the best for all contexts. Benefit-sharing systems need to be adaptive and evolve, as needed, according to the relevant situation. Some broad characteristics are, however, that they need to include accountability, equity, transparency, a broad reach of stakeholder engagement, a robust combination of tangible and intangible benefits, with benefits associated with the PA tourism from the communities and stakeholders from where they originate.

Although this special issue focuses mainly on sharing of economic benefits, there are other significant social and environmental benefits that need to be further understood and measured and future research should focus on accounting for and measuring all related benefits to give a true reflection of the range of benefits from PA tourism. Economic willingness-to-pay studies on the potential impact of introducing a compulsory tourism benefit-sharing levy on all tourism businesses in PA would provide meaningful data as to the likely acceptance of such a scheme. Another contentious issue that requires further research is the notion of who should be paying for the benefits: is it the tourist, the government or the tourism business owner, though ultimately the added cost will most likely be borne by the tourist. Future research into the willingness of tourists to bear this additional 'cost' to ensure that those living on the edge do benefit from PA tourism could assist policymakers in setting this fee and understanding the potential economic impacts of this.

This special issue of the *Journal of Sustainable Tourism* coalesces innovative approaches to benefit-sharing through new contributions to ideas surrounding empowerment (Schmidt & Uriely, 2018), measuring or accounting for the impacts (Hubert et al., 2018, this issue; Querios & Mearns, 2018; Souza et al, 2018, this issue; Snyman, 2018; Rylance et al., 2017) and understanding governance processes (Stone & Nyupane 2014). It also highlights the critical role of various stakeholder engagement in the benefit-sharing process (Querios & Mearns, 2018; Heslinga et al., 2018).

In summary, our overview and following special issue papers highlight potential areas for future research and work, including a) the multiplier effects of tourism staff spending in local economies, b) measurement methods of broader levels of benefit-sharing, c) improved reporting processes for gathering data on revenue-sharing, and d) more detailed research on the environmental, social and cultural impacts of protected area tourism in order to provide a more accurate reflection of all impacts, positive and negative. As highlighted by Strickland-Munro, Allison, and Moore (2010) and others, PA tourism involves immensely complex, dynamic systems and this, therefore, inherently requires complex, dynamic benefit-sharing systems to ensure equity and sustainability to ensure important conservation outcomes.

References

Ahebwa, W. M., van der Duim, R., & Sandbrook, C. (2012). Tourism revenue sharing policy at Bwindi Impenetrable National Park, Uganda: A policy arrangements approach. *Journal of Sustainable Tourism*, *20*(3), 377–394. doi:10.1080/09669582.2011.622768

Adams, W. M., Aveling, R., Brockington, D., Dickson, B., Elliot, J., Hutton, J., ... Wolmer, W. (2004). Biodiversity conservation and the eradication of poverty. *Science*, *12*(306), 1146–1149. doi:10.1126/science.1097920

African Impact Foundation. (2018, December 15). Conservation. Retrieved from https://africanimpactfoundation.org/conservation/.

Archabald, K., & Naughton-Treves, L. (2001). Tourism and revenue-sharing around national parks in Western Uganda: Early efforts to identify and reward local communities. *Environmental Conservation*, *28*(2), 135–149. doi:10.1017/S0376892901000145

Baghai, M., Miller, J. R. B., Blanken, L. J., Dublin, H. T., Fitzgerald, K. H., Gandiwa, P., ... Lindsey, P. (2018). Models for the collaborative management of Africa's protected areas. *Biological Conservation*, *218*, 73–82. doi:10.1016/j.biocon.2017.11.025

Balmford, A., Green, J. M. H., Anderson, M., Beresford, J., Huang, C., Naidoo, R., ... Manica, A. (2015). Walk on the wild side: estimating the global magnitude of visits to protected areas. *PLoS Biology*, *13*(2), e1002074. doi:10.1371/journal.pbio.1002074

Bango, A., & Xelelo, Z. (2017). 'Protected Areas and Community Relationships in the SubSaharan Africa: A case study of Dwesa-Cwebe Nature Reserve in the Eastern Cape, South Africa' World Congress on New Technologies, 1.

Bergstrom, R. D., & Harrington, L. M. B. (2018). Understanding agents of change in amenity gateways of the Greater Yellowstone region. *Community Development*, *49*(2), 145–160. doi:10.1080/15575330.2017.1416648

Biber-Klemm, S., & Martinez, S. (2006, January). *Access and Benefit Sharing Good practice for academic research on genetic resources. Swiss Academy of Sciences*, 1st ed., pp. 3000. Switzerland: Swiss Academy of Sciences.

Bovarnick, A., Fernandez Baca, J., Galindo, J., & Negret, H. (2010). Financial sustainability of protected areas in Latin America and the Caribbean: Investment policy guidance. Retrieved from https://www.cbd.int/financial/finplanning/g-planscorelatin-undp.pdf.

Bricker, K. S. (2001). Ecotourism development in the rural highlands of Fiji. In D. Harrison (Ed.), *Tourism and the less developed world: Issues and case studies*. Wallingford, UK: CABI, pp. 235–250.

Bricker, K. S. (2013). Ecotourism as a venue for environmental and social justice: A case study of a Fijian Vanua. In Schwab, K., and Dustin, D. (Eds.) *Just leisure: Things that we believe in* (pp. 160–166). Urbana, IL: Sagamore Publishing LLC, (2013). pp. 256.

Bruyere, B. L., Beh, A. W., & Lelengula, G. (2009). Differences in perceptions of communication, tourism benefits, and management issues in a protected area of rural Kenya. *Environmental Management*, *43*(1), 49–59. doi: 10.1007/s00267-008-9190-7

Cole, S. (2006). Information and empowerment: The keys to achieving sustainable tourism. *Journal of Sustainable Tourism*, *14*(6), 629–644. doi:10.2167/jost607.0

Convention on Biological Diversity. (2010). Strategic plan for biodiversity 2011–2020, including Aichi Biodiversity Targets. Retrieved from https://www.cbd.int/sp/.

Convention on Biological Diversity (2019). Aichi Biodiversity Targets. Retrieved from https://www.cbd.int/sp/targets/

Eagles, P. F. J., & Hillel, O. (2008). *Improving protected area finance through tourism*. Paper presented to the second meeting of the Ad Hoc Open-ended Working Group on Protected Areas, Rome, Italy. Retrieved from http://ahs.uwaterloo.ca/~eagles/documents/EaglesandHillelArticleonEconomicsandFinanceofTourisminProtectedAreas.pdf.

Emerton, L., Bishop, J., & Thomas, L. (2006). *Sustainable financing of protected areas: A global review of challenges and options*. IUCN, Gland, Switzerland and Cambridge, UK.

Franks, P., Booker, F., & Roe, D. (2018). *Understanding and assessing equity in protected area conservation: a matter of governance, rights, social impacts, and human wellbeing*. IIED Issue Paper. IIED, London.

Heslinga, J. Groote., & Vanclay, (2018). Strengthening governance processes to improve benefit sharing from tourism in protected areas by using stakeholder analysis. *Journal of Sustainable Tourism*, special issue.

Hockings, M., Stolton, S., Leverington, F., Dudley, N., & Courrau, J. (2006). *Evaluating Effectiveness: A framework for assessing management effectiveness of protected areas*. 2nd ed. IUCN, Gland, Switzerland and Cambridge, UK.

Igoe, J., Brockington, D., Randall, S., & Scholfield, K. (2008). Lessons to be learned about migration around protected areas. *Science*. www.sciencemag.org/cgi/eletters/321/5885/123?ck=nck.

Kim, Y., Kim, C., Lee, D. K., Lee, H., & Andrada, R. (2019). Quantifying nature-based tourism in protected areas in developing countries by using big social data. *Tourism Management*, *72*, 249–256. doi:10.1016/j.tourman.2018.12.005

Kothari, A. (2008). Protected areas and people: the future of the past. *Parks*, *17*(2), 23–34. Retrieved from https://www.iucn.org/sites/dev/files/import/downloads/kothari_article_parks_17_2.pdf.

Lackey, N. Q., & Bricker, K. (2019). The relationship between parks and protected areas and communities: Two scoping reviews on the current state of research globally and in North America. *The 31st Annual National Environment and Recreation Research Symposium*. April 2-6, 2019. Annapolis, Maryland, USA.

Lindsey, P. A., Miller, J. R. B., Petracca, L. S., Coad, L., Dickman, A. J., Fitzgerald, K. H., … Hunter, L. T. B. (2018). More than $1 billion needed annually to secure Africa's protected areas with lions. *PNAS*, *115*(45), E10788–E10796. published ahead of print October 22, 2018, doi:10.1073/pnas.1805048115

MacKenzie, C. A. (2012). Trenches like fence make good neighbours: Revenue sharing around Kibale National Park, Uganda. *Journal for Nature Conservation*, *20*(2), 92–100. doi:10.1016/j.jnc.2011.08.006

Makame, M. K., & Boon, E. K. (2008). Sustainable tourism and benefit-sharing in Zanzibar: The case of Kiwengwa-Pongwe Forest Reserve. *Journal of Human Ecology*, *24*(2), 93–109. doi:10.1080/09709274.2008.11906105

Mbaiwa, J. E., & Stronza, A. L. (2010). The effects of tourism development on rural livelihoods in the Okavango Delta, Botswana. *Journal of Sustainable Tourism*, *18*(5), 635–656. doi:10.1080/09669581003653500

Mutanga, C. N., Muboko, N., & Gandiwa, E. (2017). Correction: Protected area staff and local community viewpoints: A qualitative assessment of conservation relationships in Zimbabwe. *PLoS One*, *12*(9), e0184779. doi:10.1371/journal.pone.0184779

Panta, S. K., & Thapa, B. (2018). Entrepreneurship and women's empowerment in gateway communities of Bardia National Park, Nepal. *Journal of Ecotourism*, *17*(1), 20–42. doi:10.1080/14724049.2017.1299743

P2P. (2019). Benefit sharing defined. Retrieved from: http://wiki.p2pfoundation.net/Benefit_Sharing.

Plummer, R., & Fennell, D. A. (2009). Managing protected areas for sustainable tourism: prospects for adaptive co-management. *Journal of Sustainable Tourism*, *17*(2), 149–168. doi:10.1080/09669580802359301

Puhakka, R., Cottrell, S. P., & Siikamäki, P. (2014). Sustainability perspectives on Oulanka National Park, Finland: mixed methods in tourism research. *Journal of Sustainable Tourism*, *22*(3), 480–505. doi:10.1080/09669582.2013.839690

Queiros, D., & Mearns, K. (2018). Khanyayo village and Mkhambathi Nature Reserve, South Africa: a pragmatic qualitative investigation into attitudes towards a protected area. *Journal of Sustainable Tourism*, *1*. doi:10.1080/09669582.2018.1436177

Ramos, M., & Prideaux, A. B. (2014). Indigenous ecotourism in the Mayan rainforest of Palenque: empowerment issues in sustainable development. *Journal of Sustainable Tourism*, *22*(3), 461–479. doi:10.1080/09669582.2013.828730

Roe, D. & Urquhart, P. (2001). Pro-poor tourism: Harnessing the world's largest industry for the world's poor. *International Institute for Environment and Development*, in *World Summit on Sustainable Development*. UK: IIED.

Rylance, A., Snyman, S., & Spenceley, A. (2017). The contribution of tourism revenue to financing protected area management in southern Africa. *Tourism Review International, 21*(2), 139–149. doi:10.3727/154427217

Sanderson, S.E. & Redford, K. (2003). Contested relationships between biodiversity conservation and poverty alleviation. *Oryx, 37*, 389–390.

Scherl, L. M., Wilson, A., Wild, R., Blockhus, J., Franks, P., McNeely, J. A., & Mcshane, T. O. (2004). *Can protected areas contribute to poverty reduction? Opportunities and limitations.* IUCN, Gland, Switzerland and Cambridge, UK. Available from: https://www.researchgate.net/publication/237551573_Can_Protected_Areas_Contribute_to_Poverty_Reduction (accessed 30th March 2019).

Scheyvens, R. (1999). Ecotourism and the empowerment of local communities. *Tourism Management, 20*(2), 245–249. doi:10.1016/S0261-5177(98)00069-7.

Scheyvens, R. (2002). *Tourism for development: Empowering communities.* Harlow: Prentice Hall.

Schmidt, J., & Uriely, N. (2018). Tourism development and the empowerment of local communities: The case of Mitzpe Ramon, a peripheral town in the Israeli Negev Desert. *Journal of Sustainable Tourism*, 1.

Schreckenberg, K., Franks, P., Martin, A., & Lang, B. (2016). Unpacking equity for protected area conservation. *Parks, 22*(2), 11–26. Retrieved from https://ueaeprints.uea.ac.uk/61731/1/PARKS_22.2_Unpacking_Equity.pdf.

Snyman, S. (2013). Household spending patterns and the flow of ecotourism income into the villages around Liwonde National Park, Malawi. *Development Southern Africa, 30*(4-05), 640–658. doi:10.1080/0376835X.2013.832149

Snyman, S. (2016). The role of private sector ecotourism in local socio-economic development in southern Africa. *Journal of Ecotourism*, doi:10.1080/14724049.2016.1226318

Snyman, S. (2018). African tourism industry employees: expenditure patterns and comparisons with other community members. *Journal of Sustainable Tourism*, doi:10.1080/09669582.2017.1408634

Snyman, S., & Spenceley, A. (2019). *Private sector tourism in conservation areas in Africa.* CABI: Oxford.

Spenceley, A. (2014). Tourism concession guidelines for transfrontier conservation areas in SADC. Report to GIZ / SADC. Retrieved from: https://tfcaportal.org/sites/default/files/public-docs/tourism_concession_guidelines_sadc_tfcas_final.pdf

Spenceley, A., Snyman, S., & Eagles, P. (2017). Guidelines for tourism partnerships and concessions for protected areas: Generating sustainable revenues for conservation and development. Report to the CBD. Retrieved from https://www.cbd.int/tourism/.

Stone, M. T., & Nyaupane, G. (2014). Rethinking community in community-based natural resource management. *Community Development, 45*(1), 17–31. doi:10.1080/15575330.2013.844192

Strickland-Munro, J. K., Allison, H. E., & Moore, S. A. (2010). Using resilience concepts to investigate the impacts of protected area tourism on communities. *Annals of Tourism Research, 37*(2), 499–519. doi:10.1016/j.annals.2009.11.001

Strickland-Munro, J. K., & Moore, S. A. (2013). 'Indigenous involvement and benefits from tourism in protected areas: A study of Purnululu National Park and Warmun Community, Australia.'. *Journal of Sustainable Tourism, 21*(1), 26–41. doi:10.1080/09669582.2012.680466

Su, M. M., Wall, G., & Xu, K. (2016). Tourism-induced livelihood changes at Mount Sanqingshan World Heritage Site, China. *Environmental Management, 57*(5), 1024. doi:10.1007/s00267-016-0672-8

Thondhlana, G., Shackleton, S., & Muchapondwa, E. (2011). Kgalagadi Transfrontier Park and its land claimants: a pre- and post-land claim conservation and development history.). *Environmental Research Letters, 6*(2), 024009. doi:10.1088/1748-9326/6/2/024009

Tosun, C., & Timothy, D. J. (2003). Arguments for community participation in the tourism development process. *The Journal of Tourism Studies, 14*, 2–15.

Tumusiime, D. M., & Vedeld, P. (2012). False promise or false premise? Using tourism revenue sharing to promote conservation and poverty reduction in Uganda. *Conservation and Society, 10*(1), 15–28. doi:10.4103/0972-4923.92189

Umuziranenge, G., & Muhurwa, F. (2017). Ecotourism as potential conservation incentive and its impact on community development around Nyungwe National Park (NNP): Rwanda. *Imperial Journal of Interdisciplinary Research, 3*(10), 447–455.

UNEP-WCMC and IUCN (2016). *Protected Planet Report 2016.* UNEP-WCMC and IUCN: Cambridge UK and Gland, Switzerland.

Watson, J. E. M., Dudley, N., Segan, D. B., & Hockings, M. (2014). The performance and potential of protected areas. *Nature, 515*(7525), 67–73. doi:10.1038/nature13947

Wittemeyer, G., Elsen, P., Bean, W. T., Burton, C. O., & Brashares, J. S. (2008). Accelerated human population growth at protected area edges. *Science, 321*(5885), 123–126. doi:10.1126/science.1158900

Winchenbach, A., Hanna, P., & Miller, G. (2019). Rethinking decent work: The value of dignity in tourism employment. *Journal of Sustainable Tourism*, 1. doi:10.1080/09669582.2019.1566346

Woodhouse, E., Bedelian, C., Dawson, N., & Barnes, P. (2018). Social impacts of protected areas: exploring evidence of trade-offs and synergies. In Schreckenberg, K., Mace, G. & Poudyal, M. (2018). *Ecosystem services and poverty reduction: trade-offs and governance.* Routledge: London and New York. Retrieved from https://www.taylorfrancis.com/books/e/9780429016295

World Bank Group (2017). *Tourism for development: 20 reasons sustainable tourism counts for development.* Retrieved from https://openknowledge.worldbank.org/bitstream/handle/10986/28388/119954-WP-PUBLIC-SustainableTourismDevelopment.pdf?sequence=1&isAllowed=y

World Bank Group (2018). *Supporting sustainable livelihoods through wildlife tourism.* The World Bank Group Knowledge Series: Tourism for Development. Washington. Retrieved from http://documents.worldbank.org/curated/en/494211519848647950/pdf/123765-WP-REVISED-PUBLIC.pdf

Xue, L., Kerstetter, D., & Hunt, C. (2017). Tourism development and changing rural identity in China. *Annals of Tourism Research*, (*66*), 170–182. doi:10.1016/j.annals.2017.07.016

Revenue sharing from tourism in terrestrial African protected areas

Anna Spenceley, Susan Snyman and Andrew Rylance

ABSTRACT

A prerequisite for the sustainability of protected areas in Africa is the meaningful inclusion of local populations in conservation and tourism. This has been demonstrated in numerous destinations where communities receive benefits from tourism in terrestrial protected areas, they are more inclined to view it positively and conserve natural resources. This paper presents a review of revenue-sharing literature, and also an analysis of the evidence of quantified benefits accrued by local communities in Africa through institutional arrangements to share revenue or finance development projects by (1) protected areas, and (2) tourism enterprises. The review highlights the challenges of revenue sharing as well as four key components of successful revenue-sharing systems: (1) economic benefits must be clearly identified and communicated, (2) benefits are appropriate to the scale of threats to biodiversity, (3) involvement of communities in decision-making on the structure and process of the distribution system, and also how the revenues are used and (4) sufficient regulatory and institutional support is necessary to develop clear objectives, aims, goals and responsibilities. This paper constitutes the first multi-country, multi-scheme analysis of revenue sharing in terrestrial African protected areas.

Introduction

The Convention on Biological Diversity recognises that an equitable sharing of income and assets is an important component of a strategy for conservation of biodiversity (United Nations, 1992). This principle has been embraced by the international conservation community. For example, the theme of the 2003 IUCN World Parks Congress in Durban was "Benefits Beyond Boundaries," which implied an approach that married conservation goals with the interests of all affected parties, including local communities. The Durban Accord that emanated from this meeting raised the concern that, *"many costs of protected areas are borne locally – particularly by poor communities – while benefits accrue globally ..."* and urged action, *"for the benefit of protected areas so that their benefits may be conserved and equitably shared"* (IUCN, 2003, p. 1, 3). A decade later, at the 2014 IUCN World Parks Congress in Sydney, it was clear that benefit-sharing was still an area of emphasis, particularly with regards to tourism in protected areas. In their report on the congress, the IUCN's Tourism and Protected Areas Specialist Group reflected that, *"benefit-sharing from tourism and protected areas is essential in ensuring the long-term support of local communities: Without tangible, visible benefits from tourism and protected areas, it is unlikely that local communities will value and support conservation and the associated tourism. Designing and implementing effective benefit-sharing schemes is, therefore, essential to long-term conservation. Using lessons learned from past and current benefit-sharing schemes is important and should be encouraged"* (Spenceley et al., 2014, p. 9–10).

These international aims are strongly based in the scientific literature, which collectively demonstrate that a prerequisite for the sustainability of protected areas in Africa is the meaningful inclusion of local populations in conservation and tourism (Hoon, 2004; Musumali, Larsen, & Kaltenborn, 2007). For example, numerous studies have shown that where communities benefit from tourism and/or protected areas, people have more positive attitudes towards protected areas (e.g. Alexander, 2000; Currie, 2001; Gillingham & Lee, 1999; Infield, 1988; Mehta & Heinen, 2001; Sekhar, 2003; Snyman, 2014; Weladji, Moe, & Vedeld, 2003) and tourism development (Bauer, 2003; Chandralal, 2010; Lepp, 2007; Mehta & Heinen, 2001; Saarinen, 2010, Snyman, 2014).

The basic premise is that if communities receive benefits from conservation and tourism in protected areas, they will be more inclined to view it positively and conserve natural resources (e.g. Ahebwa, van der Duim, & Sandbrook, 2011; Alexander, 2000; Bauer, 2003; Chandralal, 2010; Currie, 2001; Emptaz-Collomb, 2009; Gillingham & Lee, 1999; Hulme & Murphree, 2001; Infield, 1988; Lepp, 2007; Mbaiwa, 2004a, 2004b; Mehta & Heinen, 2001; Mehta & Heinen, 2001; Sekhar, 2003; Snyman, 2014; Tumusiime & Vedeld, 2012; Wang & Pfister, 2008; Waylen, McGowan, Pawi Study Group, & Milner-Gulland, 2009; Weladji et al., 2003; Yitbarek, Tadie, Timer, & Fisher, 2013). For those community members not receiving direct benefits from tourism, the perception or belief that they could receive benefits in the future can also serve as a motivating factor in instilling positive attitudes (Walpole, Goodwin, & Ward, 2001). This anticipation of future benefits could also explain the initial enthusiasm of communities towards tourism operations (Alexander, 2000; Doxey, 1975; Sekhar, 2003). Over time, if these benefits are not realised, communities may become disillusioned and dissatisfaction may start to cause tensions between the community and the tourism operator or protected area authority (Alexander, 2000), which can impact sustainability. Benefit-sharing from tourism is just one way in which communities can gain advantages from living in proximity to conservation areas. Other ways include from access to ecosystem services through the protection of water sources, etc., access to and use of natural resources such as firewood and medicinal plants, and cultural benefits such as access to traditional or cultural sites of significance (Lockwood, Worboys, & Kothari, 2006).

Swemmer, Grant, Anneck, and Freitag-Ronaldson (2015) suggest that the term "benefit-sharing" is not well defined in the context of protected areas, and define it as " *...the process of making informed and fair trade-offs between social, economic and ecological costs and benefits within and between stakeholder groups, and between stakeholders and the natural environment...*" Swemmer et al. (2015) state that understanding stakeholders' changing perceptions of what constitutes a benefit is critical for prioritising appropriate benefits and the processes required for the sharing thereof. They propose that well-defined terminology, clear, feasible and appropriate objectives, qualitative and quantitative indicators that are developed collectively with relevant stakeholders, and appropriate and legitimate monitoring programmes can contribute greatly to the success of benefit-sharing within the conservation sector Swemmer et al., 2015, p. 17). Benefit-sharing can take many forms, including tangible, as well as intangible benefits, direct and non-direct. Tangible benefits include employment, direct income, infrastructure, etc. Intangible benefits include capacity building, skills training, cultural benefits, etc. Revenue sharing is a specific form of benefit-sharing that refers only to the revenue earned from tourism, through concessions, partnerships and other revenue-earning opportunities. Revenue sharing is, therefore, one of a suite of benefit-sharing options and should be combined with other options to maximise the benefits from tourism and protected areas.

However, Waylen et al. (2009) suggest that existing evidence does not clearly support the assumption that income from tourism changes local awareness and attitudes. Stricklund-Munro and Moore (2012) support this perspective, and suggest that economic returns are not the only way benefits are perceived and that there is a fundamental role played by culture in determining perceptions of benefits among local communities. They suggest that identifying and enhancing intrinsic, non-financial benefits related to cultural connections can assist in progressing sustainable tourism (Stricklund-Munro & Moore, 2012). Stronza and Gordillo (2008) found that non-economic benefits, such as new skills, broader experiences in managing projects and people, the exchange of ideas, expanded circles of contacts, empowerment and support for community efforts, could also influence attitudes towards conservation and tourism.

To add a further area of complexity, Swemmer et al. (2015) highlight that there is no consistent understanding of what benefit-sharing means both to protected areas and people living in the associated social ecological systems (e.g. McCool, 2012; Nkhata, Breen, & Mosimane, 2012) and state that this challenges effective implementation of "benefit-sharing" from protected areas. Therefore, to provide a clear focus for this paper, the research has specifically addressed benefit-sharing through institutional arrangements to either share revenue or finance community development projects by (1) protected areas (e.g. Ormsby & Mannle, 2006; Rylance, 2016; Spenceley, Habyalimana, Tusabe, & Mariza, 2010) or (2) tourism enterprises (e.g. Snyman, 2012, 2014, 2017; Spenceley, 2010). Therefore, it does not address other direct and indirect benefits that can be provided from tourism in protected areas, such as include employment or the sales of goods and services to tourists and tour operators, or access to natural resources within protected areas.

Revenue sharing is one form of benefit-sharing that has been promoted widely in the tourism and conservation sectors, and is characteristically driven by government, donors, NGOs and the private sector (Wynberg & Hauck, 2014). Funds may be accumulated through levies, permits and/or taxes that may be stipulated in law and then allocated to local communities (Wynberg & Hauck, 2014). This paper constitutes the first multi-country, multi-scheme analysis of revenue sharing in terrestrial African protected areas.

Research approach

Primarily, the review searched three online databases of peer-reviewed publications: Science Direct, SciVerse's Scopus and ISI Web of Science. Keywords used in searches included "tourism" AND "protected areas" OR "reserve" OR "national park" AND "benefit-sharing" OR "revenue" OR "income." Literature results contained the above keywords within their searched titles, abstracts and keywords. Second, to cross-check the publications database searches, Google Scholar was searched using the same keyword construction. The first 50 search results were compared with the papers from Science Direct, SciVerse's Scopus and ISI's Web of Science to add new publications to the reference list as well as test the completeness of the previous search process. Overall the search yielded 84 peer-reviewed papers, which were imported to Mendeley for screening. The screening process for peer reviewed literature required that the relevance of each article was verified first at a title level, and second, at an abstract level. The inclusion criteria for articles in the preliminary screening process included

- English language articles, due to the researchers' own language limitations;
- published between 1990 and present, to include recent publications; and,
- explicitly addressing the revenue sharing with local communities in Africa.

Peer reviewed literature whose full texts could not be retrieved from online references were discounted, as were duplicate papers. The remaining articles were reviewed at a full text level, with unsuitable articles rejected that did not satisfy the following inclusion criteria:

- application of research to a specific protected area and/or tourism operation as a case study;
- presentation of primary data; and,
- addressed revenue sharing according to the definition applied in this paper.

Additional grey literature (e.g. consultancy reports, occasional papers, conference and workshop presentations) that were identified by the authors were also used where it addressed the search criteria.

The review process generated a further nine papers which were reviewed alongside academic papers.

Collectively, the review generated 54 papers from 17 African countries (i.e. Botswana, Cameroon, Congo, Ethiopia, Ghana, Kenya, Lesotho, Madagascar, Mozambique, Namibia, Nigeria, Rwanda, South Africa, Tanzania, Uganda, Zambia and Zimbabwe) on tourism revenue sharing from protected areas and tourism operations within protected areas.

Results

As discussed, benefits from tourism come in many forms. From a tourism point of view, the level of commitment to benefit-sharing is determined by a number of factors such as the nature of the business, ownership and the reasons or objectives for starting such businesses (Muganda, Sirima, Moshy, & Mkumbo, 2012). For protected areas, the level of commitment to benefit, and revenue sharing is often determined by the availability of funds and political will (Rylance, Spenceley, & Spenceley, 2017). The focus in this section is specifically on revenue sharing by protected area authorities and tourism operations in Africa.

Revenue sharing by protected area authorities

Several protected area authorities in Africa have some form of revenue-sharing programme (see Table 1). These are designed to provide benefits to local communities to help offset the costs of living close to wildlife (and human–wildlife conflict), and for experiencing controls on their access to natural resources. Some programmes distribute cash to communities living around protected areas through their representative trusts (e.g. Mozambique, Tanzania), while others use the money to finance community development initiatives (e.g. Uganda, Rwanda, Tanzania, Zambia). An overview of the revenue-sharing arrangements between protected areas and local communities identified in Africa, and values of those benefits, is provided in Table 1. The proportion of revenue allocated to local communities varies from 1% (South African National Parks [SANParks]) to 20% (Uganda and Tanzania).

Uganda has provided a particular focus of research on revenue sharing (e.g. Adams & Infield, 2003; Archabald & Naughton-Treves, 2001; Ahebwa, van de Duim, & Sandbrook, 2012; MacKenzie, 2011, 2012; Sandbrook & Adams, 2012; Tumusiime & Vedeld, 2015; Vedeld et al., 2016). Respondents in Archabald and Naughton-Treves' (2001) study in Uganda listed revenue sharing as the most important advantage of living next to a national park. The majority of respondents (72%) indicated that they thought tourism revenue sharing had improved attitudes towards the protected areas, and 53% thought tourism revenue sharing was more important than sustainable use of non-timber forest products (Archabald & Naughton-Treves, 2001). By contrast, a survey of villagers around Kibale National Park in Uganda suggests that attitudes towards the park were shaped more by loss aversion than benefits received, which suggested that the protected area authority should focus more on loss mitigation than benefit provision (MacKenzie, 2012). For some protected areas in Uganda, the emphasis has shifted more to individual benefits than collective benefits (Tumusiime & Vedeld, 2012). For example, since 2006 the benefits shared from Bwindi Impenetrable National Park have been used for goat keeping, potato growing and tree planting initiatives (Tumusiime & Vedeld, 2012). The goat initiatives were supported because they provide an alternative to game meat, provide manure to fertilise crops, and so address food security and also poverty challenges (Tumusiime & Vedeld, 2012). Vedeld et al. (2016) and MacKenzie (2011) indicate that around Mt Elgon National Park and Kibale National Parks, the revenue-sharing funds are used to support schools (e.g. classroom, latrine and teacher house construction), health clinics (e.g. construction and improvement), support infrastructure (e.g. bridges, roads), livestock farming, beekeeping, coffee and tree nursery establishment, soil fertility management and water conservation, construction of elephant trenches to reduce crop raiding, and energy projects (e.g. support for construction of a biogas plant and improved cook stoves) (Moll, 2011; Petursson, 2011; UWA, 2012). Despite these individual benefits,

Table 1. Revenue sharing with local communities by protected areas in Africa.

Protected area	% tourism revenue to local communities	Financial benefits	Sources
Ghana			
Kakum National Park	20%	Unspecified	Think Ghana (2009); Bowditch (2000)
Madagascar			
National Parks	Unspecified	USD 5500 (1992/3), USD 6750 allocated to community fund spend on social projects	Durbin and Ratrimoarisaona (1996)
Mozambique			
Maputo Special Reserve	16%	USD 66,400 as cash to community trusts (2009–2013)	MSR Management Team (2013)
Rwanda			
Virunga Volcanoes National Park	5%	USD 428,000 (2005–2008) spent on social projects	Nielsen and Spenceley (2011); Spenceley et al. (2010)
South Africa			
South African National Parks (21 national parks)	1% tourism revenue and other funding partners	USD 112.3 million (FY 2014–2015: estimated). Spent on 17 socio-economic initiative projects (e.g. computer and science laboratories, nurseries, water provision, school dormitories).	SANParks (2015)
Tanzania			
Lake Manyara National Park	7.5% budget	USD 32,000 (2006–2007). Spent on teachers house, classrooms	Muganda et al. (2012)
Amani nature reserve	20%	Approx. USD 2000 per year distributed equally among 18 villages	Shoo and Songorwa (2013)
Tarangire National Park	20%	USD 329,669 (2000–2005) on community development projects (e.g. school dormitory and renovation, renovation of cattle dip)	Sachedina and Nelson (2010)
Uganda			
Bwindi Impenetrable Park	20% park entrance fees, plus USD 5 from every gorilla permit	USD 178,902 (1996–2009) Spent on local administration building, education, feeder roads, health, income generation projects, fighting crop raiding	Tumusiime and Vedeld (2012)
Kibale, Bwindi, Mgahinga National Parks	12% (1994 policy)	USD 83,000 (1995–1998). Spent on 21 schools, 4 clinics, 1 bridge, 1 road	Archabald and Naughton-Treves (2001)
Kibale National Park, Uganda	20%	USD 150,000 (1999–2009) spent on 55 projects. Average village-valued direct benefits ranged from USD 0 to USD 28,460 annually	MacKenzie (2011, 2012)
Mt Elgon National Park	20%	USD 11,313 shared between 1999 and 2002 USD 39,469 shared between 2009 (USD 6578 p/a)	Chhetri, Mugisha, and White (2003); Vedeld et al. (2016)

many community members are dissatisfied with their share of the revenues, "*A tourist pays USD 500, and the community share is to be only USD 5. This is not enough. They should give people at least USD 100. Let us compare it to the preparation of a meal: If you were a cook, would you prefer to get the smallest or the biggest portion to eat? Now, we the park neighbours are the cooks of that meal, but what do we get? The smallest portion. Is that fair?*" (Tumusiime & Vedeld, 2012, p. 20). Vedeld et al. (2016) suggest that typically direct local incomes from conservation are scarce and rather insignificant in relation to the associated costs and people's overall household incomes. Furthermore, people have been concerned that the funds are mismanaged, in part due to a lack of visibility of the funds and poorly managed projects, which are often uncompleted (MacKenzie, 2011). Despite this, a recent review suggests that the unofficial shift of Uganda Wildlife Authority towards crop raiding mitigation and income generation projects since 2008 has meant that people living closer to parks are more likely to receive the benefits (MacKenzie, Fuda, Ryaln, & Hartter, 2017).

In Rwanda, a key tourist attraction is the Mountain Gorillas in the Virunga Volcanoes National Park (Spenceley et al., 2010). Foreign non-residents pay USD 750 to hike through the park, and spend up to one hour with the endangered mammals. The park's revenue-sharing programme allocates 5% of tourism revenues to be spent on community development projects in education (e.g. constructions of schools and classrooms), environmental protection (e.g. tree planting, soil erosion control, fencing protected areas), food security (e.g. supporting beekeeping associations and basket weaving projects), and water (e.g. providing 32 25,000 litre water tanks to support 1250 people). Between 2005 and 2008, these benefits were valued at USD 428,000 (Nielsen & Spenceley, 2011; Spenceley et al., 2010). However, recent surveys around the park (Munanura, Backman, Hallo, & Powell, 2016) have found that the financial costs of crop raiding exceed the potential benefits to individuals and communities Munanura et al. (2016) state that it is critical to understand that while the benefit-sharing scheme benefits communities collectively, the cost of conservation is borne by individual households neighbouring the park. In Uganda, Baker (2005) estimated that an individual farmer adjacent to Bwindi would lose USD 472 over the 10 years of life of a banana plant due to crop raiding. Since the majority of the farmers survive on less than a dollar a day, this is a significant loss (Kidd & Giampaoli, 2006). Furthermore, households are not aware of any explicit link between community development or revenue-sharing projects to tourism and conservation in the park. Similar echoes of discontent come from Tanzania, where despite benefit-sharing of 20% of tourism revenues from the Amani nature reserve, representatives of 18 villages were discontented with their percentage share, and other benefits they received from tourism. On average, only 22.7% of surveyed households earned an income from tourism, and among these, tourism contributed 9.6% of total annual household income (Shoo & Songorwa, 2013).

Not surprisingly, the actual value of benefits accrued relates to the level of tourism receipts – so although SANParks has the lowest percentage benefit share, it appears to generate the highest value of benefit (USD 112.3 million in one year) from tourism revenues accrued from 21 national parks (SANParks, 2015).

There are also some interesting lessons from studies in Africa that describe community perceptions of benefit-sharing programmes, but without quantifying their financial impact. For example, the Liuwa National Park in Zambia, provided local community benefits from tourism including 37 school scholarships, classrooms and teacher's housing, boreholes and solar cookers (Nyirenda & Nkhata, 2013), and found that the level of success of the benefit-sharing depended on the community's fair collective bargaining strength and cooperative trust. Johannesen and Skonhoft (2005) used a bio-economic model to analyse integrated conservation and development projects in Africa. They concluded that uniform transfers to local communities as a percentage of profits (generated by entrance and hunting fees) helped to reduce poaching and increase wildlife conservation in instances where local communities' valuation of game meat was lower than the perceived monetary benefits (Johannesen & Skonhoft 2005). In Cameroon, a survey of residents in Waza National Park determined that people who benefited from the park were most in its favour, whereas the minority that disliked the park (and who did not benefit) were frustrated with management practices and promises (Bauer, 2003). In Ethiopia, a survey around Mago National Park, identified that the overall amount of revenue available to communities was actually too little to make any meaningful impact and distributed to communities on a rotational basis, and in fact park managers were not even aware of how much had been distributed (Yitbarek et al., 2013). A study undertaken in Zimbabwe was unable to identify any monetary benefits form protected areas from four protected areas: Gonarezhou National Park, Matusadona National park, Umfuruduzi Safari Area, or Cawston Ranch Safari Area; though non-monetary benefits were identified (Mutanga, Muboko, & Gandwia, 2017).

Mukanjari, Muchapondwa, Zikhali, and Bednar-Friedl (2012) suggest that a performance-linked benefit-sharing scheme can achieve socially optimal conservation. Their study on gorilla tourism in three countries (Rwanda, Uganda and Democratic Republic of Congo) suggested an alternative benefit-sharing scheme which involves the Park Agency paying the local community an amount of money directly linked to the growth in the gorilla stock (Mukanjari et al., 2012). This would provide a direct link between benefit-sharing and the associated conservation but also has a number of challenges. These include: fluctuating payments mean communities are unable to do long-term financial planning; there could be trust issues in relying on external organisations informing communities of the payment level relative to stock counts; communities might also dispute that other factors in the area may be responsible for reduced stocks and therefore communities should not be unfairly penalised, for example the potential for human to animal disease spread and poaching. Measurement and assessment of the stock could thus be problematic and undermine the equity of such a system.

Revenue sharing from tourism enterprises

Revenue sharing from tourism enterprises in Africa tends to be based on joint-venture partnership agreements, involving both a private tourism operator and a community entity. Some tour operators in Africa have made such partnerships a core element of their business model. For example, in Namibia, Wilderness Safaris (WS) has a formal joint-venture with the Torra Conservancy, a community organisation, to develop and operate the Damaraland Camp. The community were given 20% of equity annually after ten years of operation, with WS building, managing and operating the Camp. Once the community owned 100% of the camp they sold 60% back to WS and a formal joint-venture was established. The community receives a percentage of turnover from the camp, which is distributed as cash to the Conservancy board (Rylance & Spenceley, 2016; Snyman, 2012). The agreement also specifies additional benefits for local people, including preferential employment, skills training and development, and opportunity to provide goods and services to the camp (Snyman, 2012). At Rocktail Camp in South Africa, a partnership exists between WS and a Black Economic Empowerment (BEE) business. Within the BEE business, a BEE partner owns 10% and 17.5% is owned by the KwaMpukane Community Trust (Wilderness Holdings, 2016). A percentage of revenue (8.5%) is paid to the protected area authority, as well as annual turtle viewing permit fees, and the BEE business receives a fixed monthly fee from the Camp. In 2015 the community received over USD 8000 from this partnership as well as over 90% of the staff at the camp being from the local community (Wilderness Holdings, 2016). Also in South Africa, and Beyond, a private sector tourism operator negotiated a 36-year lease agreement with the Mnqobokazi and Makhasa communities for the use of Phinda Reserve, which was signed in 2006 and Beyond pay annual lease fees to their host communities to operate on the land, and paid on average USD 370,000[1] per annum between 2007 and 2013 (Rylance & Spenceley, 2016).

Groom and Harris (2008) reviewed data from two neighbouring Maasai group ranches in the wildlife corridors Amboseli and Tsavo National Parks (Kenya). They found that while one ranch generated considerable wildlife revenues from a tourist operation and community trust, the other received no direct benefits from wildlife as there was no tourist operation, no conservation trust and no other revenue streams (Groom & Harris, 2008). The overall attitude to wildlife on the ranch with wildlife revenues was significantly more positive than the other. Furthermore, they noted that attitudes within the ranch also varied significantly, depending on both the individual costs from wildlife and perceptions of the distribution of wildlife revenues (Groom & Harris, 2008, p. 242). Their analysis demonstrated that it was not the value of revenue received or the scale of costs incurred which impacted attitudes, but simply the presence or absence of wildlife benefits (Groom & Harris, 2008). Importantly, Bruyere, Beh, and Lelengula (2009) suggest that there is a need to generate consensus between protected areas and community members in Kenya on the extent to which tourism-related revenue should be distributed in the region.

An overview of the financial benefits from each of these enterprises to Community Trusts identified for tourism enterprises in protected areas in Africa is provided in Table 2. The examples demonstrate that the proportion of equity held by communities in these enterprises varies between 11.5% and 60%. As with the benefit-sharing from protected areas, the actual value of revenue received from these institutional arrangement is completely dependent on the turnover of the tourism business. The value of benefits received by communities shown in Table 2 varies widely from USD 0 per year, to USD 230,000 per year.

The community's use of their communal benefits from tourism enterprises is usually not influenced by the private sector operator. Failure of community institutions to distribute benefits in an

Table 2. Financial benefits from tourism enterprises to local communities.

Enterprise	Description of community equity/benefit	Financial benefit (USD equivalent)	Source
Botswana			
Vumbura Plains and Little Vumbura, Botswana	Partnership Wilderness Safaris and the Okavango Community Trust with a fixed annual lease fee	USD 1,292,074 (2011–2014) USD 230,600 (2015)	Snyman (2014); Wilderness Holdings (2016)
Kenya			
Koija Starbeds ecolodge, Kenya	Partnership between Koija Group Ranch and Oryx Limited. NGO support from African Wildlife Foundation	USD 108,271 given to the community in first six years. Used for healthcare, education and water provision Average annual community income between USD 20,000–30,000	Sumba, Warinwa, Lenaiyasa, and Muruthi (2007); Nthiga, Van der Duim, Visseren-Hamakers, and Lamers (2015)
Sanctuary at Ol Lentille	In 2005, a 25-year management agreement signed between Kijabe group ranch and private operator to manage lodge and conservation area at a fixed fee per year, as well as a bed-night and conservation fee for every guest at the lodge	Average annual community income is between USD 20,000 and 30,000, plus USD 250,000 on education, health and conservation	Nthiga et al. (2015)
Mozambique			
Ndzou Lodge, Mozambique	60% Mpunga community 40% Eco-MICAIA	None. Not yet profitable	Kingman (2012)
Namibia			
Damaraland Camp, Namibia	40% ownership by the Torra Conservancy	USD 74 p/p in 2004 (enough for basic groceries for 3 months) USD 790,722 (2011–2014) USD 75,000 (2015)	Spenceley (2010); Snyman (2012); Rylance and Spenceley (2016); Wilderness Holdings (2016)
Doro Nawas Camp, Namibia	45% owned by the Doro ! Nawas Conservancy	USD 69,000 (2015)	Wilderness Holdings (2016)
Rwanda			
Sabyinyo Silverback Lodge, Rwanda	Kinigi & Nyange community IGCP (NGO) Government	USD 50 per night bed fee 7.5% net sales USD 8 bed night re-payment loan	Spenceley et al. (2010); Wellard (2009)
South Africa			
Rocktail Camp, South Africa	17.5% owned by the KwaMpukane Community Trust and 10% owned by a BEE partner Fixed monthly fee paid to the BEE partner and community	USD 8000 (2015)	Wilderness Holdings (2016)
Tanzania			
Community Tourism Project Mto wa Mbu	11.5% of revenue set aside for Village Development Fund (VDF)	USD 8000 spent on village development activities (2007)	Muganda et al. (2012)

equitable, transparent manner can result in discontent in communities and no visible link being observed in the community between conservation, tourism and the benefits (Poultney & Spenceley, 2001; Snyman, 2012, 2014). The value of the benefit-sharing, and the mechanism by which it is distributed to local communities (i.e. direct cash payments, community association fund, filtered to pre-agreed projects), is usually determined by the individual negotiations between the private sector tourism operator and local community representatives. This presents a challenge for communities that have often not participated in complex contractual discussions.

Discussion

Tourism revenue sharing forms an important component of the wider benefit-sharing options for communities bordering protected areas. It is important to re-emphasise that tourism revenue sharing is only one of a suite of options of benefit-sharing from protected areas and that, in order to maximise benefits, it should be used in conjunction with other forms of benefit-sharing, including capacity building, skills training, cultural benefits and, in some cases, access to natural resources. The evidence from the analysis demonstrates that the receipt of revenue generated from tourism in protected areas in Africa has a positive influence on communities' attitudes towards conservation as well as tourism. However, simply receiving payments is not sufficient, but fraught with difficulties that can translate into negative perceptions, a number of which have been highlighted in this paper. Key elements from the findings highlight primarily that the value of revenue-sharing payments should be of sufficient volume and frequency to off-set the costs of conservation, such as crop damage, livestock depredation (Shibia, 2010), or of tourism, such as restriction of resource use (Emerton, Bishop, & Thomas, 2006). Archabald and Naughton-Treves' (2001) results suggest that if there are sufficient funds, revenue-sharing programmes may be more effective when a multi-layered approach, including both individuals suffering disproportionate costs of conservation and the general community in the definition of the target community is used. For revenue sharing from tourism enterprises, agreement with set lease fees, escalating each year, provides a community with some level of revenue growth potential (Snyman, 2012). Sinclair (2014) also indicates that it is of paramount importance that tourism initiatives generate enough revenue to change behaviour that has adverse impacts on biodiversity.

Conversely, if conservation costs are perceived to outweigh benefits from protected area tourism, then however benefits are distributed, they are unlikely to sustain widespread support for conservation (Groom & Harris, 2008; MacKenzie, 2011; Sandbrook & Adams, 2012), highlighting the complex nature of revenue-sharing schemes. In addition, benefits divided on a rotational basis between communities can mean a number of years before a community receives their next payment (MacKenzie, 2011; Yitbarek et al., 2013). This has negative impacts on the long-term financial planning by communities, and also reducing the value that each community receives. Often there is highly uneven distribution of revenues shared from tourism in protected areas, and also the costs associated with tourism and conservation, make it difficult to identify or benefit appropriate people for revenue-sharing projects (Archabald & Naughton-Treves, 2001; Spenceley, 2014). To illustrate, MacKenzie (2012) found that community members living around Kibale National Park in Uganda did not have the same levels of benefits or costs. While eight villages accrued an annual net benefit as a result of the park, 17 villages experienced a net loss. Households within 0.5 km of the park boundary accrued the highest losses from park-protected animals raiding crops and preying on livestock, while benefits were distributed up to 15 km away (MacKenzie, 2012). Similarly at Bwindi National Park, local people believed that local leaders unfairly distributed benefits, to their own villages, families or to people who bribed them (Tumusiime & Vedeld, 2012).

Beyond the direct value and perception of benefits, having a strong institutional and regulatory environment is critical for ensuring that communities receive sufficient support. The success of tasks such as developing community trusts, designating beneficiaries and choosing criteria for project selection is dependent on strong institutional support, adequate funds for the specific revenue-sharing programme, transparency and sustainability (Archabald & Naughton-Treves, 2001). Strong technical and institutional support would assist in establishing equitable and effective revenue sharing as well as developing a clear policy environment. Poorly defined policy can result in unclear institutional mandates where responsibility is unclear and often left to individuals to drive revenue-sharing programmes. High turnover rate of staff within these institutions further exacerbates this problem (Archabald & Naughton-Treves, 2001; Tumusiime & Vedeld, 2012). For example, in Mozambique the law dictates that 16% of total entrance fees within conservation areas are returned to communities. However, the distribution and frequency of payments are not clearly defined in policies, resulting in inconsistent and delayed payments to communities, creating further tension between park authorities and neighbouring communities (Rylance, 2016). Furthermore, policies that fail to address the power dynamics between park authorities and communities or are not well known and understood by communities will isolate communities from decision-making (Ahebwa et al., 2012; Stricklund-Munro, 2010). An absence of consensus within local communities can also affect their bargaining power when negotiating lease fees with private operators, resulting in sub-optimal financial benefits accruing to communities (Lapeyre, 2009).

The analysis also identified that revenue sharing may be more effective in combination with complementary projects, including sustainable use of non-timber forest products, conservation education, and problem animal control (Archabald & Naughton-Treves, 2001; Munanura et al., 2016), so that the linkages between conservation and revenue sharing are clearer. Sometimes the benefits of social infrastructure in communities (e.g. schools, water projects) or financial benefits are not directly associated with conservation, or linked to conservation responsibilities (i.e. not poaching) (Munanura et al., 2016; Spenceley, 2014; Yitbarek et al., 2013). For example, in Ethiopia challenges to revenue-sharing programmes have included information provision, imbalanced roles and responsibilities; compromised accountability; and the lack of connection between revenue and wildlife in the minds of recipients (Yitbarek et al., 2013). Munanura et al. (2016) found in their analysis that revenue-sharing schemes in general had several structural constraints such as political influence (for example, self-serving interests and corruption by local political leaders), poor conservation linkages and limited participation of the most socially and economically disadvantaged living near the Park.

Communicating the benefits provided (and potential benefits), so that there is greater awareness of what has been provided and to whom, may be more important than actual benefit delivery in determining the long-term viability of private sector tourism operator/community relations (Munanura et al., 2016; Snyman, 2013). This is often dependent on the capacity and transparency of the protected area authority to design and implement a benefit-sharing system (whether from tourism or other forms of benefit-sharing) that mitigates against the previously-identified challenges. In many areas there are no formal systems to manage benefit-sharing programmes, or they are not functioning correctly (Makame & Boon, 2008). Ormsby and Mannle (2006) stress the importance of park representatives clearly communicating potential park benefits and drawbacks to communities in order to share information as well as foster positive interactions.

Based on the literature review and lessons from other revenue-sharing programmes in Africa included in this analysis, incorporating the following four components of successful revenue-sharing programmes should be considered by protected area authorities and private sector businesses aiming to enter into joint-venture agreements or other partnerships with local communities in Africa,

(1) economic incentives that provide immediate benefits to local people must be clearly identified and communicated (Bookbinder, Dinerstein, Rijal, Cauley, & Rajouria, 1998);
(2) incentives must be appropriate to the scale of current threats to biodiversity (Dinerstein, Rijal, Bookbinder, Kattel, & Rajuria, 1999);
(3) local people should be involved in decision-making for both developing the structure and the process of distribution system as well as the utilisation of the tourism sharing revenues (Snyman, 2013; Tumusiime & Vedeld 2012); and,
(4) sufficient regulatory, technical and institutional support is critical to ensure that revenue-sharing programmes have clearly allocated objectives, aims, goals and responsibilities.

Conclusion

This paper has aimed to conduct the first multi-country, multi-scheme analysis of tourism revenue sharing in Africa's terrestrial protected areas. Whilst internationally there is strong support for deriving community benefits from tourism, only a limited proportion of the literature has been able to quantify the direct revenue from tourism shared with communities by protected areas agencies and tourism enterprises. It has highlighted that direct revenue payments to communities are not significant, especially when considered on a per capita basis. Nevertheless, sharing revenue from tourism with local communities is a necessary factor in building local support for protected areas and conservation, but it is not sufficient. The complexity in determining what type and level of revenue generated and how it accrues to neighbouring communities are also important considerations. Nkhata et al. (2012, p. 59) confirms that benefit-sharing is founded on complex long-term social exchanges that involve multiple interest groups with divergent expectations and experiences. There is also a growing recognition that revenue-sharing initiatives, as a form of benefit-sharing, cannot be successful in the absence of well-functioning governmental organisations, local community institutions and free-market mechanisms (Schuklenk & Kleinsmidt, 2006 and Wunder, 2007 in Nyirenda & Nkhata, 2013). Understanding all the factors which impact revenue sharing is crucial for proper long-term planning, management and sustainability of tourism development while ensuring mutual benefits to business operators, visitors and the local community (Muganda et al., 2012). Despite these challenges, Tumusiime and Vedeld (2012) do not suggest that tourism revenue sharing is abandoned as an approach, but recommend that the shortcomings be addressed. They note, *"The problem is not with tourism revenue sharing as an idea and as a concept, but with the difficulties in putting it into real-world practice"* (Tumusiime & Vedeld, 2012, p. 26).

It is clear that where communities have limited involvement in the conservation of wildlife, and a negligible share of the benefits of wildlife tourism, but suffer inconveniences of living with wildlife, there is conflict. In these areas, wildlife is not considered a priority form of land use since it does not offer immediate benefits to the local communities (Sibanda & Omwega, 1996). Revenue sharing should, therefore, be included as one of a suite of benefit-sharing options, allowing for a diversification of risk as well as the potential to have a broader impact in a community. There is no doubt that revenue sharing is a complex process, requiring the interaction of numerous stakeholders and various conditions in order to have the potential to make a long-term sustainable difference socio-economically and in terms of biodiversity conservation. Future research focusing on whether additional economic benefits, such as employment and supply chain opportunities have a greater per capita impact, would be useful avenue of investigation.

Note

1. USD 2.6 million in lease fees paid from 2007 to 2013.

Disclosure statement

No potential conflict of interest was reported by the authors.

References

Adams, W. M., & Infield, M. (2003). Who is on the Gorilla's payroll? Claims on tourist revenue from a Ugandan National Park. *World Development, 31*(1), 177–190.

Ahebwa, W. M., van der Duim, R., & Sandbrook, C. (2011). Tourism revenue sharing policy at Bwindi Impenetrable National Park, Uganda: A policy arrangements approach. *Journal of Sustainable Tourism, 20*, 377–394.

Ahebwa, W. M., van de Duim, R., & Sandbrook, C. (2012). Tourism revenue sharing policy at Bwindi Impenetrable National Park, Uganda: A policy arrangements. *Journal of Sustainable Tourism, 20*(3), 377–394.

Alexander, S. (2000). Resident attitudes toward conservation and black howler monkeys in Belize: The community baboon sanctuary. *Environmental Conservation, 27*(4), 341–350.

Archabald, K., & Naughton-Treves, L. (2001). Tourism revenue sharing around national parks in Western Uganda: Early efforts to identify and reward local communities. *Environmental Conservation, 2*, 135–149.

Baker, J. (2005). *Mountain gorillas: Crop raiding and conflict at BINP*, Uganda. Canterbury: Durrell Institute of Conservation and Ecology, University of Kent.

Bauer, H. (2003). Local perceptions of Waza National Park, Northern Cameroon. *Environmental conservation, 30*(2), 175–181.

Bookbinder, M. P., Dinerstein, E., Rijal, A., Cauley, H., & Rajouria, A. (1998). Ecotourism's support of biodiversity conservation. *Conservation Biology, 12*(6), 1399–1404.

Bowditch, N. H. (2000). *Ghana tourism revenue sharing action plan* (Report to USAID Mission to Ghana). Sigma One Corporation. Retrieved from http://pdf.usaid.gov/pdf_docs/Pnacu164.pdf

Bruyere, B., Beh, A., & Lelengula, G. (2009). Differences in perceptions of communication, tourism benefits, and management issues in a protected area of Rural Kenya. *Environmental management, 43*(1), 49–59. doi:10.1007/s00267-008-9190-7

Chandralal, K. (2010). Impacts of tourism and community attitudes towards tourism: A case study in Sri Lanka. *South Asian Journal of Tourism and Heritage, 3*(2), 41–49.

Chhetri, P., Mugisha, A., & White, S. (2003). Community resource use in Kibale and Mt Elgon National Parks, Uganda. *Parks, 13*(1), 28–37.

Currie, J. (2001). *Towards sustainable conservation: Community perceptions of game reserves in the Maputaland region of South Africa: Comparative case studies of Phinda and Mkhuze game reserves* (Master's thesis). Middlesex University, London.

Dinerstein, E., Rijal, A., Bookbinder, M., Kattel, B., & Rajuria, A. (1999). Tigers as neighbors: Efforts to promote local guardianship of endangered species in lowland Nepal. In J. Seidensticker, P. Jackson, & S. Christie (Eds.), *Riding the tiger: Conserving the tiger in a human-dominated landscape* (pp. 316–333). Cambridge: Cambridge University Press.

Doxey, G. V. (1975). A causation theory of visitor-resident irritants: Methodology and research inferences. In Proceedings of the Travel Research Association 6th Annual Conference (pp. 195–198). San Diego, CA: Travel Research Association.

Durbin, J. C., & Ratrimoarisaona, S.-N. (1996). Can tourism make a major contribution to the conservation of protected areas in Madagascar? *Biodiversity & Conservation, 5*(3), 345–353.

Emerton, L., Bishop, J., & Thomas, L. (2006). *Sustainable financing of protected areas. a global review of challenges and options*. Gland; Cambridge: IUCN.

Emptaz-Collomb, J. (2009). *Linking tourism, human wellbeing and conservation in the Caprivi Strip, Namibia*. Florida, FL: University of Florida.

Gillingham, S., & Lee, P. (1999). The impact of wildlife-related benefits on the conservation attitudes of local people around the Selous Game Reserve, Tanzania. *Environmental Conservation, 26*(3), 218–228.

Groom, R., & Harris, S. (2008). Conservation on community lands: The importance of equitable revenue sharing. *Environmental Conservation, 35*(3), 242–251.

Hoon, P. (2004). Impersonal markets and personal communities? Wildlife, conservation and development in Botswana. *Journal of International Wildlife Law & Policy, 7*(3), 143–160.

Hulme, D., & Murphree, M. (Eds.). (2001). *African wildlife & livelihoods: The promise and performance of community conservation*. Portsmouth: Heinemann, Curry.

Infield, M. (1988). Attitudes of a rural community towards conservation and a local conservation area in Natal, South Africa. *Biological Conservation, 45*(1), 21–46.

Johannesen, A. B., & Skonhoft, A. (2005). Tourism, poaching and wildlife conservation: What can integrated conservation and development projects accomplish. *Resource and Energy Economics, 27*, 208–226.

Kidd, C., & Giampaoli, P. (2006). *Consultation with Batwa communities on their land rights under the Ugandan State: A joint consultation for the United Organisation for Batwa Development in Uganda (UOBDU) and the Uganda Land Alliance.* cited in Tumusiime, D. M. and Vedeld, P. (2012). op. cit.

Kingman, A. (2012). *Concessions and sustainability in Chimanimani: NDZOU Camp & MICAIA, Ndzou Lodge*. Presentation at the "Concessioning Tourism Opportunities in Conservation Areas and Maximising Rural Development" Workshop, Maputo, Mozambique.

Lapeyre, R. (2009). Revenue sharing in community – Private sector lodges in Namibia. *Tourism Economics, 15*(3), 653–669.

Lepp, A. (2007). Resident's attitudes toward tourism in Bigodi village, Uganda. *Tourism Management, 28*, 876–885.

Lockwood, M., Worboys, G. L., & Kothari, A. (2006). *Managing protected areas: A GLOBAL GUIDE*. London: IUCN, Earthscan.

MacKenzie, C. A. (2011). Trenches like fences make good neighbours: Revenue sharing around Kibale National Park, Uganda. *Journal for Nature Conservation, 20*, 92–100.

MacKenzie, C. A. (2012). Accruing benefit or loss from a protected area: Location matters, *Ecological Economics, 76*, 119–129.

MacKenzie, C. A., Fuda, R. K., Ryaln, S. J., & Hartter, J. (2017). Drilling through conservation policy: Oil exploration in Murchison Falls Protected Area, Uganda. *Conservation and Society, 15*(3), 322–333.

Makame, M. K., & Boon, E. K. (2008). Sustianable tourism and benefit-sharing in Zanzibar: The case of Kiewenga-Pongwe Forest Reserve. *Journal of Human Ecology, 24*(2), 93–108.

Maputo Special Reserve Management Team. (2013). *Development of Maputo special reserve*. Maputo: Presentation to World Bank Mission.

Mbaiwa, J. (2004a). The socio-cultural impacts of tourism development in the Okavango Delta, Botswana. *Journal of Tourism and Cultural Change, 2*(3), 163–184.

Mbaiwa, J. (2004b). The success and sustainability of community-based natural resource management in the Okavango Delta, Botswana. *South African Geographical Journal, 86*(1), 44–53.

McCool, S. F. (2012). *Distributing the benefits of nature's bounty: A social justice perspective*. Paper presented at International Symposium on Managing Benefit Sharing in Changing Social Ecological Systems, Windhoek, Namibia.

Mehta, J., & Heinen, J. (2001). Does community-based conservation shape favourable attitudes among locals? An empirical study from Nepal. *Environmental Management, 28*(2), 165–177. doi:10.1007/s002670010215

Moll, R. (2011). *The privilege to cooperate: Assessing the impact of collaborative resource management agreements on local communities' livelihoods around Mount Elgon National Park, Uganda* (MA thesis). Department of Economics, University of Oslo, Norway. cited in Vedeld et al (2016). op. cit.

Muganda, M., Sirima, A., Moshy, B., & Mkumbo, P. (2012). Sharing tourism benefits with the local community: A business perspective from the grassroots in Tanzania. *Ethiopian Journal of Environmental Studies and Management, 5*(3), 240–251.

Mukanjari, S., Muchapondwa, E., Zikhali, P., & Bednar-Friedl, B. (2012). Evaluating the prospects of benefit sharing schemes in protecting mountain gorillas in Central Africa. *Environment for Development Discussion Paper* EfD DP 12–16.

Munanura, I., Backman, K., Hallo, J., & Powell, R. (2016). Perceptions of tourism revenue sharing impacts on Volcanoes National Park, Rwanda: A sustainable livelihoods framework. *Journal of Sustainable Tourism, 24*(12), 1709–1726.

Musumali, M., Larsen, T., & Kaltenborn, B. (2007). An impasse in community based natural resource management implementation: The case of Zambia and Botswana. *Oryx, 41*(3), 306–313.

Mutanga, C. N., Muboko, N., & Gandiwa, E. (2017). Protected area staff and local community viewpoints: A qualitative assessment of conservation relationships in Zimbabwe. *PLoS One, 12*(5), e0177153. Retrieved from https://doi.org/10.1371/journal. pone.0177153.

Nielsen, H., & Spenceley, A. (2011). The success of tourism in Rwanda: Gorillas and more. In P. Chunhan-Pole & M. Angwafo (Eds.), *Yes Africa Can: Success stories from a dynamic continent* (231–249). Washington, DC: World Bank.

Nkhata, B. A., Breen, C., & Mosimane, A. (2012). Engaging common property theory: Implications for benefit sharing research in developing countries. *International Journal of the Commons, 6*(1), 52–69.

Nthiga, R. W., Van der Duim, R., Visseren-Hamakers, I. J., & Lamers, M. (2015). Tourism-conservation enterprises for community livelihoods and biodiversity conservation in Kenya. *Development Southern Africa, 32*(3), 407–423. doi:10.1080/0376835X.2015.1016217

Nyirenda, V. R., & Nkhata, B. A. (2013). Collaborative governance and benefit-sharing in Liuwa Plain National Park, Western Zambia. *Parks, 19.1*, 103–114.

Ormsby, A., & Mannle, K. (2006). Ecotourism benefits and the role of local guides at Masoala National Park, Madagascar. *Journal of Sustainable Tourism, 14*(3), 271–287.

Petursson, J. G. (2011). *Institutions and transboundary protected area management: The case of Mount Elgon, Uganda and Kenya* (PhD dissertation). Norwegian University of Life Sciences, Ås.

Poultney, C., & Spenceley, A. (2001). *Practical strategies for pro-poor tourism, Wilderness Safaris South Africa: Rocktail Bay and Ndumu Lodge* (Pro-Poor Tourism Working Paper No. 1). London: Overseas Development Institute.

Rylance, A., & Spenceley, A. (2016). Applying inclusive business approaches to nature-based tourism in Namibia and South Africa. *Tourism, 64*(4), 371–383.

Rylance, A. (2016). Estimating tourism's contribution to conservation area financing in Mozambique. *Tourism and Hospitality Research, 17*(1), 24–33.

Rylance, A., Snyman, S., & Spenceley, A. (2017). The contribution of tourism revenue to financing protected area management in southern Africa. *Tourism Review International, 21*, 139–149.

Saarinen, J. (2010). Local tourism awareness: Community views in local tourism awareness: Community views in Katutura and King Nehale. *Development Southern Africa, 27*(5), 713–714.

Sachedina, H., & Nelson, F. (2010). Protected areas and community incentives in savannah ecosystems: A case study of Tanzania's Massai Steppe. *Oryx, 44*(3), 390–398.

Sandbrook, C., & Adams, W. M. (2012). Accessing the impenetrable: The nature and distribution of tourism benefits at a Ugandan National Park. *Society & Natural Resources, 25*(9), 915–932.

SANParks. (2015). *Annual Report 2014/2015*. Pretoria: South African National Parks.

Schuklenk, U., & Kleinsmidt, A. (2006). North south benefit sharing arrangements in bioprospecting and genetic research: A critical ethical and legal analysis. *Developing World Bioethics, 6*(3), 122–134. doi:10.1111/j.1471-8847.2006.00149.x

Sekhar, N. (2003). Local people's attitudes towards conservation and wildlife tourism around Sariska Tiger Reserve, India. *Journal of Environmental Management, 69*, 339–347.

Shibia, M. G. (2010). Determinants of attitudes and perceptions on resource use and management of Marsabit National Reserve, Kenya. *Journal of Human Ecology, 30*(1), 55–62.

Shoo, R. A., & Songorwa, A. N. (2013). Contribution of eco-tourism to nature conservation and improvement of livelihoods around Amani nature reserve, Tanzania. *Journal of Ecotourism, 12*(2), 75–89.

Sibanda, B. M. C., & Omwega, A. K. (1996). Some reflections on conservation, sustainable development and equitable sharing of benefits from wildlife in Africa: The case of Kenya and Zimbabwe. *South African Journal of Wildlife Research, 26*(4), 175–181.

Sinclair, J. (2014). Volunteers add vitality and energy to Fraser Island management Fraser Island Defenders Organisation. IUCN World Parks Congress, Sydney, 12–19 November. cited in Spenceley, A. (2016). Tourism and protected areas: Comparing the 2003 and 2014 IUCN World Parks Congress, *Tourism and Hospitality Research*, 1-16. doi:10.1177/1467358415612515.

Snyman, S. (2012). Ecotourism joint ventures between the private sector and communities: An updated analysis of the Torra Conservancy and Damaraland Camp partnership, Namibia. *Tourism Management Perspectives, 4*, 127–135.

Snyman, S. (2013). *High-end ecotourism and rural communities in southern Africa: A socio economic analysis* (Unpublished PhD thesis). University of Cape Town, South Africa.

Snyman, S. (2014). Partnerships between private sector ecotourism operators and local community in the Okavango Delta, Botswana: The case of the Okavango Community Trust and Wilderness Safaris. *Journal of Ecotourism, 13*(2–3), 110–127.

Snyman, S. (2017). The role of private sector ecotourism in local socio-economic development in southern Africa. *Journal of Ecotourism, 16*(3), 247–268. doi:10.1080/14724049.2016.1226318.

Spenceley, A. (2010). *Tourism product development interventions and best practices in sub-Saharan Africa*. Part 1: Synthesis report, Part 2: Case studies. Report to the World Bank, 27 December 2010.

Spenceley, A. (2014). *Benefit sharing from natural heritage: Examples and challenges from Africa*. Keynote, Inkasa Research Symposium, Cape Town Hollow Hotel, South Africa.

Spenceley, A., Habyalimana, S., Tusabe, R., & Mariza, D. (2010). Benefits to the poor from gorilla touirsm in Rwanda. *Development Southern Africa, 27*(5), 647–662.

Spenceley, A., Hvenegaard, G., Mader, R., Snyman, S., Nikolaeva, E., Bushell, R. & 18 TAPAS Group Members. (2014). IUCN WCPA Annual report to the steering committee: Tourism and Protected Areas Specialist Group (TAPAS Group), Reporting period 1 January 2014–31 December 2014.

Stricklund-Munro, J., & Moore, S. (2012). Indigenous involvement and benefits from tourism in protected areas: A study of Purnululu National Park and Warmun Community, Australia. *Journal of Sustainable Tourism, 21*(1), 26–41. doi:10.1080/09669582.2012.680466.

Stricklund-Munro, J. K. (2010). *Understanding the interactions among local communities, protected areas and tourism: Case studies of Kruger National Park and Purnululu National Park* (Master's thesis). Australia: Murdoch University.

Stronza, A., & Gordillo, J. (2008). Community views of ecotourism. *Annals of Tourism Research, 35*(20), 448–468.

Sumba, D., Warinwa, F., Lenaiyasa, P., & Muruthi, P. (2007). *The koija starbeds ecolodge: A case study of a conservation enterprise in Kenya*. (AWF Working Papers). Nairobi: African Wildlife Foundation.

Swemmer, L., Grant, R., Annecke, W., & Frietag-Ronaldson, S. (2015). Toward more effective benefit-sharing in South African National parks. *Society and Natural Resources, 28*(1), 4–20.

Think Ghana. (2009). Kakum Park's revenue sharing formula to be revised. Retrieved from http://tourism.thinkghana.com/pages/news/200907/29702.php

Tumusiime, D. M., & Vedeld, P. (2012). False promise or false premise? Using tourism revenue sharing to promote conservation and poverty reduction in Uganda. *Conservation and Society, 10*(1), 15–28.

Tumusiime, D. M., & Vedeld, P. (2015). Can biodiversity conservation benefit local people? Costs and benefits at a strict protected area in Uganda. *Journal of Sustainable Forestry, 34*(8), 761–786.

Uganda Wildlife Authority (UWA). (2012). *Revenue sharing policy and guidelines*. Kampala: Uganda Wildlife Authority.

United Nations. (1992). *Convention on biological diversity*. Geneva: United Nations Environment Programme (UNEP).

Vedeld, P., Cavanagh, C., Petursson, J. G., Nakakaawa, C., Moll, R., & Sjaastad, E. (2016). The political economy of conservation at mount elgon, Uganda: Between local deprivation, regional sustainability, and global public goods. *Conservation Society, 14*, 183–194.

Walpole, M., Goodwin, H., & Ward, K. (2001). Pricing policy for tourism in protected areas: Lessons from Komodo National Park, Indonesia. *Conservation Biology, 15*, 218–227.

Wang, Y. A., & Pfister, R. E. (2008). Residents' attitudes toward tourism and perceived personal benefits in a rural community. *Journal of Travel Research, 47*(1), 84–93.

Waylen, K., McGowan, P., Pawi Study Group, & Milner-Gulland, E. (2009). Ecotourism positively affects awareness and attitudes but not conservation behaviours: A case study at Grand Riviere, Trinidad. *Oryx, 43*(3), 343–351.

Weladji, R., Moe, S., & Vedeld, P. (2003). Stakeholder attitudes toward wildlife policy and the Bénoué Wildlife Conservation Area, North Cameroon. *Environmental Conservation, 30*(4), 334–343.

Wellard, M. (2009). *Sabyinyo silverback lodge: A community partnership for conservation.* Kampala: International Gorilla Conservation Program

Wilderness Holdings. (2016). Wilderness integrated annual report 2016. Retrieved from http://www.wilderness-group.com/system/assets/188/original/Integrated_Annual_Report_2016.pdf?1482302236

World Conservation Union (IUCN). (2003). World Parks Congress. Retrieved from www.iucn.org/wpc2003, cited in Staiff, R., and Bushell, R. (2004) Tourism and protected areas: Benefits beyond boundaries, Conference report, *Annals of tourism research*, 31 (3), pp.723-726.

Wynberg, R., & Hauck, M. (2014). People, power, and the coast: A conceptual framework for understanding and implementing benefit sharing. *Ecology and Society, 19*(1), 27. https://doi.org/10.5751/ES-06250-190127

Yitbarek, T. W., Tadie, D., Timer, G., & Fisher, A. (2013). Evaluating governance processes in the sharing of revenues from wildlife tourism and hunting in Ethiopia. *Environmental Conservation, 40*(3), 253–265.

Economic impacts of tourism in protected areas of Brazil

Thiago do Val Simardi Beraldo Souza, Brijesh Thapa ,
Camila Gonçalves de Oliveira Rodrigues and Denise Imori

ABSTRACT
Protected areas (PAs) are globally considered as a key strategy for biodiversity conservation and provision of ecosystem services. Economic impact analyses generate quantifiable estimates of the tourism interdependencies which provide a tangible understanding of effects on local and regional economies. Federal PAs of Brazil are the object of the research based on the Money Generation Model2 (MGM2) methodology along with modifications to address particular issues in a developing country context. The significance of this research is to assist managers, inform policy-makers, conservation and commercial stakeholders, local communities, and the public at large of the value that PAs serve for conservation as well as engines for benefit-sharing. Results identified that each dollar Brazil invested in the PA system produced $7 in economic benefits. Also, the total economic contributions generated more than $1.3 billion in total sales, $342 million in personal income, $473 million in value added to the GDP, and supported 43,602 jobs nationally. The study reinforced the view that economic impacts of tourism influences PAs as well as indirectly to other businesses and the local communities.

Introduction

The creation of protected areas (PA) is an effective strategy for biodiversity conservation and provision of ecosystem services. PAs offer a spectrum of outdoor recreation opportunities that attract visitors as well as assist to raise societal awareness and support for conservation (Viveiros de Castro, Souza, & Thapa, 2015). Globally, the number of visits to terrestrial PAs is estimated to be 8 billion per year and they generate approximately $600 billion[1] in direct in-country expenditures. Such economic impacts are considerably positive, and it has been argued that it could be even greater if $10 billion per year were applied to safeguard the PAs (Balmford et al., 2015). The importance of tourism to generate revenues for management has been advocated, but yet, conservation priorities have consistently experienced budgetary constraints (Eagles, 2002; Thapa, 2013). In addition, since tourism is interdependent with other stakeholders (e.g. businesses, services, governments and local communities), visitor expenditures that permeate within and outside PAs stimulate job creation and entrepreneurial activities (e.g.: tour operations, hospitality business, etc.). Moreover, as most PAs are located in rural and remote regions, the economic impacts of visitor expenditures aid in the creation of alternative income and poverty alleviation sources for local communities (Emerton, Bishop, & Thomas, 2006; Ferraro & Merlin, 2014).

Consequently, the economic, social and environmental benefits and costs of can ultimately affect entire regions. Hence, examination of the positive and negative impacts is warranted in order to mitigate costs and formulate sustainable tourism initiatives within a PA and adjacent communities. Nevertheless, within the tripod of sustainability, the economic perspectives have increasingly witnessed interest among policy-makers and decision-makers with regards to investment decisions at PAs (Balmford et al., 2015). As a result, more countries are developing visitor expenditure analyses at site and system levels: USA (Cullinane & Koontz, 2016), Canada (The Outspan Group, 2011), Australia (Driml, 2010), Finland (Huhtala, Kajala, & Vatanen, 2010), Namibia (Turpie, Barnes, Lange, & Martin, 2010), South Africa (Saayman, Rossouw, & Saayman, 2010), and Brazil (Medeiros & Young, 2011).

Furthermore, the positive economic impacts based on visitations and expenditures are also used to increase public opinion and support for PAs. For example, since 1969, the U.S. National Park Service (NPS) has formally estimated the economic contributions of outdoor recreation – 140 million visits generated $6.4 billion in spending and $4.7 billion in personal income (Stynes, 2001). The visitation continued its increase with 307 million visits in 2015 along with $16.9 billion in spending. Based on the data, the estimated contribution to the national economy was $32 billion in output and $11.1 billion in labor income, and created 295,000 jobs (Cullinane & Koontz, 2016). Such periodically formulated estimations have been instrumental for NPS to advocate for fiscal support and negotiate budget allocations through the years.

Other national agencies have also followed the same strategy. Parks Canada documented 20 million recreation visits in 2008–2009, and reported the economic significance of $2.24 billion, with 41,720 full-time jobs and tax revenues of $163 million (The Outspan Group, 2011). Furthermore, even developing countries such as Namibia have also evaluated the economic importance of their PA system. Results estimated that 180,000 recreation park visits were recorded with an overall expenditure of $176 million. The direct contribution to the GDP was $80 million along with a total contribution of $147 million[2] (Turpie et al., 2010).

In Brazil, the first attempt to evaluate the economic impacts of tourism in the National PAs system was conducted in 2010 via the Money Generation Model2 (MGM2; Medeiros & Young, 2011). Based on visitation numbers in 2009 (3.8 million visits), the impact was estimated to be approximately between $144 and $163 million in tourist spending[3] (see Table 1). Such estimates reflect the importance to the economy since the PAs organizational entity's (Chico Mendes Institute for Biodiversity Conservation [ICMBio]) budget spent for the management of the whole system was only $69 million in 2010 (ICMBio, 2011). Subsequently, Semeia (2014) used an indirect approach based on demographic census analysis to evaluate the economic potential of tourism in PAs. Results identified that investments in infrastructure and services in the PAs at all levels (national, state, and local) would result in an increase of 55,800 jobs and $1.6 billion in total sales supported by tourism in the gateway communities. Furthermore,

Table 1. Comparative analysis between Brazil, USA and Canada.

Country	Brazil[a]	USA[b]	Canada[c]
No. of PAs considered	310	360	129
Visitor numbers	3.8 mi	307 mi	20 mi
Budget	U$ 69 mi	U$ 2.6 bi	U$ 545 mi (2015)
Visitor revenues	U$ 6.2 mi	U$ 430 mi	U$ 90 mi
Revenues/% of the budget	(9%)	(17%)	(16%) (2015)
Economic significance	U$ 144/163 mi	U$ 32 bi	U$ 2.5 bi
Visitor spending	–	U$ 16.9 bi	U$ 2.0 bi
Added value	–	U$ 18.4 bi	–
Labor income	–	U$ 11.1 bi	U$ 1.44 bi
Jobs	–	295,000	41,720
Year	2010	2015	2008/2009

[a] Medeiros and Young (2011).
[b] Cullinane and Koontz (2016).
[c] The Outspan Group (2011).

Muanis, Serrão, and Geluda (2009) estimated that national PAs would need $209 million from Brazilian government in investments to consolidate infrastructure in all federal system.

The ICMBio is the federal agency responsible for the management of the Federal PAs in Brazil. ICMBio manages a system of 325 federal PAs in a total of 79 million hectares (ICMBio, 2016a). Despite the enormous size of the PAs system along with its important biodiversity, the associated budget has not been fully substantiated in Brazil. Additionally, it is still ambiguous with respect to the impact and value added of tourism via visitors' spending due to lack of empirical research. Therefore, the purpose of this research was to estimate the economic impacts of tourism in the federal system of PAs of Brazil. More specifically, the following objectives were formulated and examined:

(1) Assess local expenditures of visitors in select PAs.
(2) Examine the current income, value added, and jobs generated by each PA.
(3) Estimate the current income, value added, and jobs generated by the entire PA system.
(4) Estimate the income, value added, and jobs generated by the entire PA system in a forecasted improved scenario.

To accomplish the objectives, this study adapted a framework methodology based on the MGM2 (Stynes, Propst, Chang, & Sun, 2000) along with minor modifications to address contextual issues for developing countries. Economic impact analyses generate quantifiable estimates of the tourism interdependencies which provide a tangible aspect of the industry effects, and ultimately the importance of the services provided by the PAs to the region and country (Stynes et al., 2000). The results of this research can assist in decision-making for managers and policy-makers, inform conservation and commercial stakeholders, and local communities, as well as the general public about the economic values and subsequent conservation importance of PAs.

Economic impact analysis

Economic impact is part of a group of analyses that can be used to evaluate tourism in PAs (other analyses include: fiscal impact analysis, financial analysis, demand analysis, benefit cost analysis, feasibility study and environmental impact assessment) (Stynes, 1997). Since each type of analysis is somewhat specific, the problems to be addressed may require more than one method. For example, an economic impact evaluation may request a prior demand analysis to forecast potential increase in volume of tourism activity.

Economic impact analysis describes the interrelationships between economic sectors, and creates estimates of the possible changes in a certain economy due to actual or future scenarios. For example, visitors spend money in PAs and gateway communities, and their expenditures create and support local economic activity. Economies are interconnected systems where producers and consumers interact (Cullinane & Koontz, 2016). Raw material processed by producers and manufacturers is combined with inputs from other economic sectors to generate output. This, in turn, is the input for yet other economic sectors which add value to the intermediate goods and generate their own output until the goods and services reach the final consumer. Consequently, if demand for a product changes, a ripple effect will be experienced at multiple levels and sectors of an economy (Cullinane & Koontz, 2016). This impact is described in terms of direct effects, indirect effects, and induced effects of the initial consumer spending:

- Direct effects are the changes caused by visitor spending in the businesses that sell directly to visitors (i.e. lodges, campgrounds, restaurants, grocery stores, etc.). However, for the trade-sector businesses (i.e. grocery and sporting goods stores, fuel stations, etc.), the direct output effects need to consider only the capture rate, which is the percentage of spending that stays in the region by first round sellers (retailers), local wholesalers, as well as manufacturers/producers (Crompton, 2010; Stynes, 2001).

- Indirect effects are the changes generated when lodges and other directly affected businesses buy goods and services from other businesses within the local region (Crompton, 2010). For service-sector businesses, the indirect output effect includes all items purchased and fixed operating expenses associated with operating a business (e.g. office supplies, electricity, maintenance, etc.). For trade-sector businesses, indirect output includes only the fixed operating expenses, and not the cost of items purchased for resale.
- Induced effects are the changes generated through household spending of personal income received directly or indirectly from tourist spending. For example, spending by employees of tourist lodges on meals, gas, etc. supports additional jobs in non-tourism businesses, and therefore allows additional rounds of local spending across a broad range of economic sectors (Cullinane & Koontz, 2016).

The secondary effect is the sum of the indirect plus the induced effects. The sum of direct, indirect, and induced effects is termed the total effects of visitor spending (Crompton, 2010). These interactions between consumers and producers are described by economic input–output (I-O) models that capture the effects of visitor spending by regional economic multipliers. To measure economic effects, it is necessary to obtain multipliers and ratios from I-O models. I-Os are matrices that describe the interdependencies and flows of money between different sectors within a certain economy (Cunha et al., 2008). I-O captures how much each industry demands from all the others. It also presents the proportion of sales that is used to pay salaries income, benefits, proprietors' income and taxes, as well as the number of employees in each sector. The basic calculations for the I-O matrix, coefficients, effects and multipliers are noted in the following (Guilhoto, 2011).

The multiplication of visitor expenditures by regional multipliers is used to calculate the economic effects of tourist spending to regional economies. There are two possible metrics to present economic effects of tourism in PAs: economic impacts and economic contributions (Cullinane & Koontz, 2016). Economic contribution analysis captures the gross economic activity generated in the regional economy by the total number of visitors in the PA. Economic contribution studies include expenditures by local visitors (the ones who live in the gateway community) and non-local visitors (the ones who come from outside). Economic impact analysis reports the economic net changes in the regional economy generated by the new money brought in to the local economy by visitors from outside the community. Economic impacts exclude spending by local visitors based on the assumption that if they decide not to visit the PA, they would spend in another recreation activity within the local community.

Furthermore, economic significance analysis is more geared to measure the relative importance and magnitude of the total contribution of tourism in PAs to regional economies. However, economic impact analysis is more indicated to measure the inflow of visitors and spending in a local community generated by non-local visitors (Crompton, 2010). Economic impacts are estimated by the following equation (Stynes et al., 2000):

Economic effects $=$ number of visitors $\times$ average spending per visitor $\times$ economic multipliers.

The variables in the equation require the following information:

(1) Collect or estimate the number of tourists that visit the PAs and surrounding area.
(2) Estimate average spending per visitor in the region.
(3) Apply economic multipliers to measure the ripple effects of the visitor expenditures.

The three inputs on the right side of the equation are derived in several ways, dependent on resources available for the study and degree of desired accuracy. The NPS in the USA developed a model called "Money Generation Model2" (MGM2) as the first attempt to estimate the impacts of visitation in the PAs. This initial approach only presented aggregated data (U.S. Dept. of Interior, National Park Service [USDI-NPS], 1990). Following MGM2, two applications were available for tourism economic research: Regional Input–Output Modeling System (RIMS II) from the US Bureau of Economic

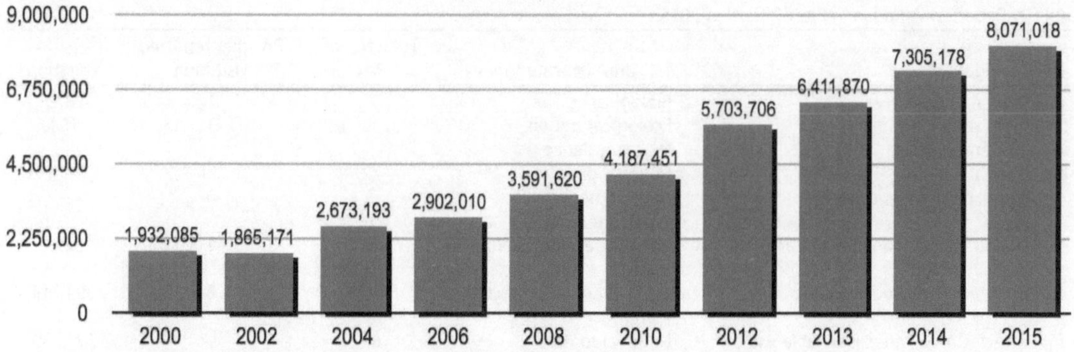

Figure 1. Total visitation/year in protected areas of Brazil.

Analysis (U.S. Dept. of Commerce, Bureau of Economic Analysis [USDC BEA], 1992), and Micro-Implan Recreation Economic Impact Estimation System (MI-REC/IMPLAM) (Stynes & Propst, 1992, 1996). Both systems provided improvements with the option to desegregate spending per industry sector and visitor segments in application that could be manipulated; however, both required the purchase of multipliers. Stynes et al. (2000) developed the second generation of the MGM called MGM2 for the NPS, which also offered desegregated spending sectors and option for visitor segments. MGM2 was released as an open version (non-proprietary) for state and local agencies with the option to use with generic multipliers (Chang, 2001). This required no acquisition of specific local multipliers, hence facilitated and reduced the cost of analysis, and likely made MGM2 as the preferred approach in economic analysis in other countries too, such as Australia, Finland and Brazil (Buultjens & Luckie, 2004; Cui, Mahoney, & Herbowicz, 2013; Huhtala et al., 2010; Medeiros & Young, 2011). There are two other systems that should also be noted: Travel Economic Impact Model developed by the US Travel Data Center (1997), and the Tourism Satellite Account by the World Travel and Tourism Council (WTTC, 1996), both of which that are more focused to estimate national impact levels.

Methods

MGM2 methodology (Stynes et al., 2000) was chosen to calculate the economic contributions of the PAs to the Brazilian economy. This method requires three inputs: number of visitors, visitor expenditures, and multipliers, all of which were collected from different sources.

Number of visitors

PAs in Brazil report visitor arrival statistics annually to the central office of ICMBio. Since 2000, visitation has grown from 1.9 million to more than 8 million in 2015 (see Figure 1) (ICMBio, 2016a). Visitation is allowed in most types of PAs (all of them if environmental education is considered) (ICMBio, 2016a) (Table 2). From the 8 million visitors in 2015, national parks and forests received 93% of total. The research collected data from 58 national parks and 36 national forests managed by the federal agency ICMBio. Of the 94 PAs that responded the survey, 62 reported visitation in 2015 (Souza, 2016).

The study considered two scenarios: (1) actual reported visitation for 2015, and (2) improved scenario of 13.4 million visitors computed based on a demand analysis.

First scenario

The Brazilian PAs were distributed in recreation opportunity classes to develop visitor segments (see Souza, 2016). The recreation opportunities are derived from activities in different settings and have three different attributes: Physical, Social, and Managerial. The combinations of these attributes create classes or zones where recreation opportunities may occur. The classes vary from Primitive Use

Table 2. Reported visitation in PAs of Brazil in 2015.

IUCN categories	Brazilian categories	Total No. of PAs	PAs that reported visitation	Visitation
(1a) Strict nature reserve	Biological reserve	31	1	2375
	Ecological station	32	1	154
(1b) Wilderness area	No equivalence			–
(2) National park	National park	72	38	7,149,112
(3) Natural monument or feature	Natural monument	3	0	–
	Wildlife refuge	7	0	–
(4) Habitat/species management area	Relevant ecological interest area	16	1	3294
(5) Protected landscape/seascape	Environmental protected area	33	3	394,744
(6) Protected area with sustainable use of natural resources	National forest	67	17	371,339
	Extractives reserve	62	1	150,000
	Fauna reserve	0		–
	Sustainable development reserve	2		–
Final total		325	62	8,071,018

Note: Correspondences based on a Ministry of Environment Report (Ministério do Meio Ambiente [MMA], 2007).
IUCN: International Union for Conservation of Nature.
Source: ICMBio (2016b).

where PAs offer few or no recreation options, facilities or services. The Primitive PAs are generally located in remote locations (e.g. Jatuarana National Forest in Amazonia) with difficult access and low demand. Conversely, Highly Intensive PAs offer a great variety of attractions, facilities, and services. They are located close to state capitals or large cities with high population density, easy access and high usage rates (e.g. Tijuca National Park in Rio de Janeiro). Between the spectra, there are other classes such as Semi-Primitive, Extensive and Intensive.[4] Since most visitations are located in PAs of Extensive, Intensive and Highly Intensive Use, this study focused the collection of visitors' expenditures from these three classes (see Figure 2).

Second scenario

A prediction of the visitation potential of the PA system based on the recreation classes was used (see Souza, 2016). Basically, the indicators from the physical, social and managerial attributes were correlated to the number of visitors at the respective PA. The results were used to perform an estimation for the entire PA system of 325 units in a scenario where all PAs would offer basic facilities and services based on their class of recreation use. Results illustrated that the entire federal system could likely receive substantially more visitors – 13.4 million instead of the 8 million reported in 2015.

Visitor expenditures

The second input for the economic analysis – average spending per visitor – was collected via two different techniques: on-site interviews and emails (see Crompton, 2010) during January–February 2016. The study focused on visitors at three PAs from the most visited based on recreation class use (Extensive, Intensive, and Highly Intensive), with the purpose to develop an average visitor spending for each class (Driml & McLennan, 2010; Huhtala et al., 2010). The three selected PAs were: São Franciso de Paula National Forest, Chapada dos Guimarães National Park and Tijuca National Park.

São Francisco de Paula National Forest (*Extensive Use*) – the manager provided an email list of 365 visitors. Based on email contact requests and subsequent reminders, a total of 131 visitors completed the questionnaire (28% response rate). Of which, 108 were deemed useable. Chapada dos Guimarães National Park (*Intensive Use*) – the manager provided an email list of 4134 visitors. Similarly, based on email contact and reminders, 329 visitors responded (8% response rate), while 229 were useable.

Figure 2. Recreation opportunity classes of use in national parks and forests of Brazil. Source: Souza (2016).

Tijuca National Park (*Highly Intensive Use*) – for this park, visitors were interviewed on-site using a systematic sampling process (Crompton, 1999) at two different access points. Data collection occurred during the week and weekends for a period of 30 days. A total of 116 responses were collected, and 97 were included in the study.

The questionnaire was one-page in length with only essential data requested for this research. Due to field logistics and response issues, the questionnaire was developed in a digital (Qualtrics software) and a standard paper version in English and Portuguese languages. The survey collected origin (ZIP code) to identify internationals, locals, and non-local residents. The relevant questions included length of stay, party size of the group, and the amounts spent in specific categories: accommodation, meals, gas and oil, local transportation, retail stores, activities and guided tours and other expenses. In addition, questions were also included to identify multiple destination visitors.

Visitor expenditures need to be carefully analyzed as it can be a potential source of error (Crompton, 2010). Hence, researchers (Cook, 2013; Crompton, 2010; Cullinane & Koontz, 2016; Huhtala et al., 2010) have enumerated several issues that need to be performed to derive an accurate participation of visitors for the analysis: definition of visitor and visit, proper spending region, definition of local visitors and zero spending visitors and multiple destination trips, which were considered in the data analyzes.

Multipliers

National ratios and multipliers were derived from the I-O Table of Brazil 2013 – 68 sectors (Guilhoto, 2015). The premise is that the national economy did not have significant changes between the year of the I-O table (2013) and the year of the visitor survey (2015). A closed matrix inclusive of families' expenditures was developed to calculate the Type II multipliers. Type II multipliers were used to calculate secondary effects (indirect and induced).

Following the MGM2 methodology, this study focused mainly on the direct effects with greater attention to income and value added as the more reliable impact measurements. The direct effects were presented by spending categories. Impacts were presented in terms of sales, value added, personal income and jobs (Stynes et al., 2000):

(1) Sales are the sales of business within the region to visitors after considering the capture rate.
(2) Jobs correspond to the number of jobs supported by tourist expenditures. Job effects consider full-time, part-time and seasonal jobs.
(3) Personal income refers to proprietor's income, salary income and wages.
(4) Value added is a contribution measurement of a region or industry to the gross product, national or state. Rents and profits, personal income and indirect business taxes summed are included in Value Added. It corresponds to the final price of the product or service after removal of the costs of production minus non-labor.

Direct effects

Multipliers and economic ratios were multiplied to each sales category from the local economy. Specific multipliers for five economic sectors were developed, in order to captures the economic differences of each industry (Crompton, 2010). Direct effects are estimated for each category of expenditure with the following formulas (Stynes, 2001):

$$\begin{aligned}\text{Direct jobs} &= (\text{direct sales} \times \text{capture rate}) \times \text{jobs/sales ratio}\\ \text{Direct personal income} &= (\text{direct sales} \times \text{capture rate}) \times \text{personal income/sales ratio}\\ \text{Direct value added} &= (\text{direct sales} \times \text{capture rate}) \times \text{value added/sales ratio}\end{aligned}$$

Total effects

The procedure for total effects is similar to direct effects via the Type II multipliers for each sector. They represent the variation in sales, income, jobs and value added for each additional unit of sales. Indirect effects are estimated in an aggregate form with the following formulas

(Stynes et al., 2000):

$$\begin{aligned}\text{Total sales} &= (\text{direct sales} \times \text{capture rate}) \times \text{Type II sales multiplier}\\ \text{Total jobs} &= (\text{direct sales} \times \text{capture rate}) \times \text{total jobs/sales ratio}\\ \text{Total personal income} &= (\text{direct sales} \times \text{capture rate}) \times \text{total income/sales ratio}\\ \text{Total value added} &= (\text{direct sales} \times \text{capture rate}) \times \text{total value added/sales ratio}\end{aligned}$$

National multipliers

Multipliers and ratios were developed for the Brazilian economy from the I-O Table of 2013 – 68 sectors (Guilhoto, 2015). The data were formulated from National Accounts based on the methodologies described in Guilhoto and Sesso Filho (2005) and Guilhoto and Sesso Filho (2010) (see Table 3).

The sectors from the I-O matrix used to extract the multipliers correspond to the spending categories in the visitor survey (see Table 4). The table contents also show the equivalent International Standard Industrial Classification (ISIC) Revision 4 and the Brazilian CNAE (Código Nacional de Atividades Econômicas). Creative, arts and entertainment activities sector was the closest sector to extract Organized Activities multipliers due to its similar classification under the ISIC section R—Art, Entertainment and Recreation. The category Other Expenses used the average from all sectors.

Results

Visitors were divided into two segments for each recreation class: local and non-local. Sufficient data were not available to statistically define an international segment. On average, based on visitor spending survey data, 20% were local and 80% were non-local. Also, locals spent one day at the PA while non-locals spent three and four days in the region. From the sampled 94 PAs (national parks and forests), a total of 7,273,684 recreation visits were allocated to the three recreation opportunity classes based on these two visitor segments (see Figure 3). The same proportion was kept to include visitors in other PA categories to calculate total economic contributions of the entire system.

Visitor expenditures

The questionnaire collected information about expenditures inside the PAs and within the region. The expenditures were divided by number of days in the region, and per group size if responses were on behalf of an entire group. Then, based on how important (1–10), the PA was for their decision to visit the region, only the percentage informed by the visitor was considered as final spending per visit per day. The expenditure averages were organized per segment (recreation class and local/non-local) on a visit per day basis (see Table 5). On average, local visitors spent $10.42 per visit, and non-local visitors accounted for $50.74. Local visitors spent more money on meals, local transportation, and activities

Table 3. National multipliers and ratios developed from the Input–Output (I-O) Table for Brazil (2013) – 68 sectors.

	Direct effects ratios				Total effects multipliers (II)			
Sector	Jobs/MM sales	Personal Income/sales	Value added/sales	Sales II	Jobs II/MM sales	Income II/ sales	VA II/ sales	Sales I
Accommodation	61.59	0.45	0.58	4.50	145.69	1.21	1.70	1.62
Eating and drinking	76.78	0.41	0.50	4.63	165.54	1.20	1.64	1.79
Creative, arts and entertainment activities	94.31	0.50	0.57	4.71	178.85	1.31	1.78	1.59
Terrestrial transport	40.16	0.35	0.45	4.58	117.10	1.10	1.52	1.97
Wholesale trade and retail trade, except motor vehicles	60.74	0.46	0.64	4.36	138.47	1.20	1.71	1.53
Average	55.60	0.36	0.46	3.80	124.27	1.00	1.39	1.42

Source: Souza (2016) spreadsheet attached.

Table 4. Corresponding I-O table sectors to the spending categories.

Categories in the questionnaire	Corresponding sectors IOM	Corresponding sectors CNAE
Gas and oil	Wholesale trade and retail trade, except motor vehicles	46/47 – Wholesale trade and retail trade, except motor vehicles
Local transportation	Terrestrial transport	49 – Terrestrial transport
Retail stores	Wholesale trade and retail trade, except motor vehicles	46/47 – Wholesale trade and retail trade, except motor vehicles
Meals	Food processing	56 – Food and beverage service activities
Accommodation	Accommodation	55 – Accommodation
Activities and guided tours	Creative, arts and recreation activities	90 – Creative, arts and recreation activities
Other expenses	Average	–

and guided tours. Non-locals spent more on accommodation, meals, gas and oil, activities, and guided tours. It should be noticed that the visitor expenditures were collected after the travel based on post-visitor memory, which often results in lower than actual levels of expenditure.

Tourism economic contributions and impacts – national effects

National direct and total contributions of visitor expenditure attributed to PAs

National direct contributions consider the direct effects of the expenditures attributable to the days spent in the PAs for all visitors. It also considers a capture rate of 100% with the assumption that all goods are produced nationally. Based on this notion, visitors' expenditures in PAs generated more than $347 million in Direct Sales, $153 million in Personal Income, $194 million in Value Added to the GDP, and supported 23,813 direct jobs. These values consider only local expenditures without considering transportation from the origin to the destination. Furthermore, national total contributions consider the direct plus indirect plus induced effects. Hence, visitors' expenditures generated more than $1.3 billion in total sales, $342 million in personal income, $473 million in value added to the GDP, and supported 43,602 jobs nationally.

Based on recreation class of use, the Extensive classification represented 24% of visitation and 14% of the direct effects. PAs with Intensive Use noted 34% of visitation and 51% of the direct effects. The Highly Intensive classification was responsible for 39% of the visitation and 35% of the direct effects. Due to different expenditures in spending categories, Highly Intensive PAs had more direct effects, but Intensive Use had more total direct effects (see Table 6).

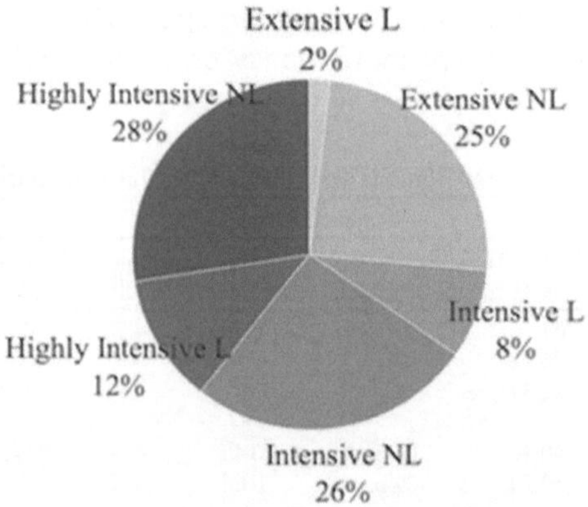

Figure 3. Percentage of visitors in each recreation class segment (L-local, NL-non-local).

Table 5. Average expenditures per recreation class.

	Recreation classes (L – local, NL – non-local)					
	Extensive		Intensive		Highly Intensive	
Spending categories	L	NL	L	NL	L	NL
Accommodation	$0.32	$5.39	$0.00	$18.49	$0.00	$15.65
Meals	$0.63	$8.01	$6.06	$12.33	$3.97	$11.86
Gas and oil	$0.00	$7.26	$5.33	$9.09	$0.16	$12.52
Local transportation	$0.32	$1.36	$0.35	$5.99	$6.28	$6.91
Activities and guided tours	$0.00	$0.69	$6.37	$14.70	$0.50	$12.52
Retail stores	$0.00	$0.82	$0.47	$2.90	$0.16	$3.75
Other expenses	$0.00	$0.38	$0.35	$0.82	$0.00	$0.76
Total expenditure per tourist/day	**$1.26**	**$23.91**	**$18.93**	**$64.32**	**$11.07**	**$63.97**

Only the total expenditure per tourist/day are in bold. They are the sum of column.

National total contributions attributed to PAs in the improved scenario

Based on the aforementioned improved scenario, the results in total effects would result in more than $2.8 billion in Total Sales Captured, US $752 million in Personal Income, US $1 billion in Value Added to the GDP, and support 94,180 jobs for the country (see Table 7).

Discussion

The results contribute academically to the economic impacts analysis literature as it relates to developing countries. This study collected primary visitor expenditures data in three different PAs to develop generic expenditures profiles for the most expressive recreation classes with local and non-local visitor segments. Additionally, based on the 2013 I-O matrix of the Brazilian economy, eight specific multipliers and ratios for each spending category were developed for the national level. Using this approach, it was possible to estimate not only sales, but also value added, income and jobs. Overall, this methodology improved the reliability of the results based on primary data collection and rigor of analysis (Stynes et al., 2000).

The economic impacts identified were higher than the estimations from Medeiros and Young (2011). Instead of the $144/$163 million in total sales generated via spending by 3.8 million visitors in 2010 (Medeiros & Young, 2011), this study found that the 8 million visitors in 2015 economically contributed towards $1.2 billion in total sales captured. It should be noted that the ICMBio Central Office provided visitor numbers for both studies, and the variance on total sales are due to differences on expenditures and multipliers. In addition, Medeiros and Young (2011) used total expenditures and separated visitors in three spending profiles between $12 and $34, while the current study used seven categories (i.e. accommodation, meals, gas and oil, local transportation, activities, retail stores and other expenses) and separated PAs visitors in six profiles between $1 and $64. The use of more visitor spending profiles and spending categories offered a more comprehensive account with regard to the capability to further analyze

Table 6. Economic contributions of visitor spending to the national economy.

Spending category	Sales captured	Jobs	Personal income	Value added
Accommodation	$84,310,383	5,193	$37,939,672	$48,900,022
Meals	$76,042,342	5,838	$31,177,360	$38,021,170
Gas and oil	$64,984,842	3,947	$29,893,027	$41,590,299
Local transportation	$36,875,676	1,481	$12,906,486	$16,594,054
Activities and guided tours	$64,745,257	6,106	$32,372,628	$36,904,796
Retail stores	$16,523,036	1,004	$7,600,596	$10,574,743
Other expenses	$4,384,808	244	$1,585,838	$2,002,395
Total direct effects	**$347,866,345**	**23,813**	**$153,475,610**	**$194,587,482**
Secondary effects	$946,109,299	19,789	$189,397,361	$279,103,778
Total effects	**$1,293,975,644**	**43,602**	**$342,872,971**	**$473,691,260**

Only the total expenditure per tourist/day are in bold. They are the sum of column.

Table 7. Economic contributions of visitor spending to the national economy: improved scenario.

Spending category	Sales captured	Jobs	Personal income	Value added
Direct effects	$621,680,792	42,725	$274,734,880	$347,625,218
Total effects	$2,821,185,706	94,180	$752,935,141	$1,045,872,259

economic impacts. Moreover, total expenditure differences (i.e. maximum values between $34 and $64) illustrated a variance in the market prices between the years of 2010 and 2015 that affected results. Finally, the other difference is that Medeiros and Young (2011) used generic multipliers adopted from the US economy, while this study developed the multipliers for each spending category based from an I-O matrix of the Brazilian economy. This further allowed to capture greater details in spending effects.

Nevertheless, both studies reinforce the idea that PAs are engines of economic development as expenses in conservation and recreation are actually investments to create and strengthen jobs, income, and gross domestic product (GDP). The total sales capture of $1.3 billion demonstrated to be almost eight times the ICMBio budget of $182 million in 2015 (ICMBio, 2016c). In essence, for every $1 that Brazil invests in the PA system, it generates $7 in economic benefits nationally. Moreover, it has also been noted that investments of approximately $220 million for infrastructure development in all federal PAs, as estimated by Muanis et al. (2009) would increase visitation to 13.4 million, and would generate another $1 billion in total value added. In fact, this potential economic impact is one-third of the $2.9 GDP of Brazilian State of Roraima during the same year.

The limitations of the estimates rely on the precision of the three inputs: number of visits, spending averages, and multipliers (Cullinane & Koontz, 2016). With respect to PAs visitation numbers, this data was based on the information provided via ICMBio headquarters. Basically, the total number of visitors compiled is based on counting procedures that vary from entry ticket sales, visitor counts at the entry gates, visitors signature book, estimations, appointments, and information from the tourist companies. Among such data collection methods, each PA uses one or more to compile the total visitors count as it is dependent on the settings and demand level. However, it is recommended that ICMBio could develop guidelines to increase scientific rigor in data collection, and define a standardized count procedure of visits and visitors. The importance of this is emphasized since the most significant input is estimate of visits, followed by the average spending figures, and distribution of visitors across segments (Stynes et al., 2000).

About expenditures, one limitation is that spending profiles were derived from just three PAs that represented the different recreation classes. Further studies should increase the number of PAs surveyed. Among the three surveyed PAs, visitor expenditure was subject to 6.5% +/− sampling error on average. Future research should focus on increasing the sample size of PAs as well as visitors' segments to include international tourists. It is also recommended to collect data during different times of the year, such as the high and low season in order to capture any potential differences. Interpretation of the results should consider that the survey conducted in Tijuca National Park was applied only during two months of summer. Chapada dos Guimarães National Park and São Francisco National Forest had list of emails that were collected over two years. In order to strengthen the visitor spending data and avoid inflation on total expenditures, the recommended potential sources of errors were treated with the differentiation of local visitors vs. non-local, zero spending visitors, local region and multi-purpose trip (Crompton, 2010). It should also be noticed that the visitor expenditures were collected after the trip based on post-visitor memory, which often results in lower than actual levels of expenditure.

Future research can continue to monitor economic impacts over the years for the federal PAs of Brazil. However, the prerequisite input variable – number of visitors – will need to be collected annually, since expenditures and multipliers can be used for a few years (Huhtala et al., 2010). Since this study only analyzed the effects at the national level, it is recommended that development of local multipliers and the economic impact evaluations be conducted at local and state levels too. Hence, the next phase of this study is to further explore and examine at the state and local levels based on the data already compiled and calculations of local and state level multipliers.

Conclusion

The study revealed the economic magnitude of an important ecosystem service provided by PAs – tourism and outdoor recreation. Visitation at PAs demonstrated to be an impactful mechanism to develop local economies and the tourism industry in Brazil as each dollar invested in management generates $7 for the economy. The total economic contributions generated more than $1.3 billion in total sales, $342 million in personal income, $473 million in value added to the GDP, and supported 43,602 jobs nationally. The study reinforced that economic impacts of tourism influence the PAs as well as indirectly to other businesses and the local communities. Since a good proportion of the sampled PAs are located in remote areas, the economic effects can further generate greater economic benefits for local communities that have a higher household dependency on the surrounding natural resources. However, investments in PAs are a necessity to ensure conservation of the ecosystem services and quality of visitors' experiences. These results highlighted the importance of tourism in PAs and the adjacent regions for the Brazilian economy, and encourage increase in budget allocation for conservation. Such additional investments would stimulate the local economies and generate benefits for communities to ensure the development of sustainable destinations.

Notes

1. Monetary values are presented in US Dollars.
2. 1.00 US Dollar equals to 13.90 Namibian Dollars (9 September 2016).
3. 1.00 US Dollar = 3.17 Brazilian Real (15 August 2016) Conversion used for the entire document.
4. For more information about the recreation use classification, see Souza (2016).

Acknowledgments

We wish to thank all colleagues, analysts and managers of national parks and forests who completed the questionnaires of PA attributes. In particular, to Edenice Brandão, manager of São Francisco de Paula National Forest, Cintia Brazão from Chapada dos Guimarães National Park, and Ernesto Vivieros de Castro from Tijuca National Park. We also wish to thank Brian Child and Lori Pennington-Gray for the strong incentives for the development of a economic analysis with primary data about tourism in protected areas of Brazil.

Disclosure statement

The findings and conclusions in this article are those of the authors and do not necessarily reflect the views of the Chico Mendes Institute of Biodiversity Conservation.

Funding

Coordination for the Improvement of Higher Education Personnel (CAPES) - Science Without Borders Program.

ORCID

Brijesh Thapa http://orcid.org/0000-0002-0163-5580

References

Balmford, A., Green, J. M. H., Anderson, M., Beresford, J., Huang, C., Naidoo, R. ... Manica, A. (2015). Walk on the wild side: Estimating the global magnitude of visits to protected areas. *PLoS Biology, 13*(2), e1002074. doi:10.1371/journal.pbio.1002074

Buultjens, J., & Luckie, K. (2004). *Economic impact of selected national parks in North-Eastern New South Wales. 88 s.* Sustainable Tourism Cooperative Research Centre. Retrieved from http://sustain.pata.org/wp-content/uploads/2015/02/Buultjens31021_NPNSW.pdf (October 28 2016).

Cunha, F. S.C., Takasago, M., Guilhoto, J., Rocha De Farias, A., Imori, D., Mollo, M. L. R., de Andrade, J. (2008). O Turismo E a Economia Brasileira: Uma Discussão Da Matriz De Insumo-Produto [Tourism and the Brazilian economy: A discussion about the input-output matrix]. *Proceedings of the 14° Congresso da APDR*. Tomar: Instituto Politecnico de Tomar. Retrieved from SSRN: http://ssrn.com/abstract=2408051

Chang, W. (2001). *Variations in multipliers and related economic ratios for recreation and tourism impact analysis* (Dissertation thesis). Michigan State University, East Lansing, MI. p. 150.

Chico Mendes Institute for Biodiversity Conservation (ICMBio). (2011). Relatorio de Gestão 2010 [Management report 2010]. Retrieved from http://www.icmbio.gov.br/acessoainformacao/images/stories/relatorios/Relatorio_de_gestao_2010_INTERNET.pdf

Chico Mendes Institute for Biodiversity Conservation (ICMBio). (2016a). *Visitors numbers in protected areas* [internal document]. Brasilia: Coordination of Public Use and Business CGEUP.

Chico Mendes Institute for Biodiversity Conservation (ICMBio). (2016b). Area das Unidades de Conservação Federais por categoria [Size of protected areas per category]. Retrieved from http://www.icmbio.gov.br/portal/images/stories/servicos/geoprocessamento/DCOL/dados_tabulares/Area_UC_federais_categoria_agosto_2016.pdf

Chico Mendes Institute for Biodiversity Conservation (ICMBio). (2016c). Relatório de Gestão 2015 [Management report 2015]. Retrieved from http://www.icmbio.gov.br/acessoainformacao/images/stories/relatorios/Relatorio_de_gestao_2015_INTERNET.pdf

Cook, P. S. (2013). *Impacts of visitor spending on the local economy: Mesa Verde National Park, 2012* (Natural Resource Report No. NPS/NRSS/EQD/NRR—2013/667). Fort Collins, CO: National Park Service.

Crompton, J. (1999). *Measuring the economic impact of visitors to sports tournaments and special events.* Ashburn, VA: National Recreation and Park Association. (Research Series).

Crompton, J. (2010). *Measuring the economic impact of park and recreation services.* Ashburn, VA: National Recreation and Park Association. (Research Series).

Cui, Y., Mahoney, E., & Herbowicz, T. (2013). *Economic benefits to local communities from national park visitation, 2011* (Natural Resource Report No. NPS/NRSS/EQD/NRTR—2013/631). Fort Collins, CO: National Park Service.

Cullinane T. C., & Koontz, L. (2016). *2015 National Park visitor spending effects: Economic contributions to local communities, states, and the nation* (Natural Resource Report No. NPS/NRSS/EQD/NRR—2016/1200). Fort Collins, CO: National Park Service.

Driml, S. (2010). *The economic value of tourism to national parks and protected areas in Australia.* Gold Coast: CRC for Sustainable Tourism.

Driml, S., & McLennan, C. (2010). *Handbook on measuring the economic value of tourism to national parks.* Gold Coast: CRC for Sustainable Tourism.

Eagles, P. J. (2002). Trends in park tourism: Economics, finance and management. *Journal of Sustainable Tourism, 10*(2), 132–153.

Emerton, L., Bishop, J., & Thomas, L. (2006). *Sustainable financing of protected areas: A global review of challenges and options.* Gland: IUCN.

Ferraro, P. J., & Merlin, H. (2014). Quantifying causal mechanisms to determine how protected areas affect poverty through changes in ecosystem services and infrastructure. *Proceedings of the National Academy of Sciences, 111*, 4332–4337.

Guilhoto, J. J. M. (2011). *Análise de insumo-produto: Teoria e fundamentos. livro em elaboração* [Input-Output analysis: Theory and fundamentals. Book in press]. Versão Revisada: Departamento de Economia. FEA-USP.

Guilhoto, J. J. M. (2015). *Input-output table of Brazil 2013 - 68 sectors.* São Paulo: NEREUS - Núcleo de Economia Regional e Urbana da Universidade de São Paulo.

Guilhoto, J. J. M., & Sesso Filho, U. A. (2005). Estimação da Matriz Insumo-Produto a Partir de Dados Preliminares das Contas Nacionais [Estimating the input-output matrix from preliminary data of national accounts]. *Economia Aplicada, 9*(2), 277–299.

Guilhoto, J. J. M., & Sesso Filho, U. A. (2010). Estimação da Matriz Insumo-Produto Utilizando Dados Preliminares das Contas Nacionais: Aplicação e Análise de Indicadores Econômicos para o Brasil em 2005 [Estimating the input-output matrix from preliminary data of national accounts: Application and analysis of economic indicators for Brazil in 2005]. *Economia & Tecnologia, 23.*

Huhtala, M., Kajala, L., & Vatanen, E. (2010). Local economic impacts of national park visitors' spending: The development process of an estimation method. (978-951-40-2224-1). Retrieved from http://www.metla./julkaisut/workingpapers/2010/mwp149.htm

Medeiros, R., & Young, C. E. F. (2011). *Contribuição das unidades de conservação brasileiras para a economia nacional: Relatório Final* (p. 120). Brasília: UNEP—WCMC.

Ministério do Meio Ambiente (MMA). (2007). *Informe nacional sobre áreas protegidas do Brasil.* Série Áreas Protegidas do Brasil. Brasília, DF: Author.

Muanis, M., Serrão, M., & Geluda, L. (2009). *Quanto custa uma unidade de conservação federal?: Uma visão estratégica para o financiamento do sistema nacional de unidades de conservação (Snuc)* [How much does a federal protected area cost?: A strategic vision for the financing of the national system of protected areas (Snuc)]. Rio de Janeiro: Funbio.

Saayman, M., Rossouw, R., & Saayman, A. (2010). *The socio-economic impact of the Kruger National Park.* Potchefstroom: Institute for Tourism and Leisure Studies, North—West University.

Semeia. (2014). *Unidades de conservação no Brasil: A contribuição do uso público para o desenvolvimento socioeconômico/ Instituto Semeia* [Protected areas in Brazil: The contribution of public use to socioeconomic development/] (p. 53). São Paulo: Semeia Institute.

Souza, T. V. S. B. (2016). *Recreation classification, tourism demand and economic impact analyses of the federal protected areas of Brazil* (Unpublished doctoral dissertation). University of Florida, Gainesville, FL.

Stynes, D. J. (1997). *Economic impacts of tourism: A handbook for tourism professionals. Illinois bureau of tourism.* East Lansing, MI: Tourism Research Laboratory, Michigan State University.

Stynes, D. (2001). *Economic impact concepts.* Retrieved from http://www.msu.edu/course/prr/840/econimpact/concepts.htm

Stynes, D. J., & Propst, D. B. (1992). A system for estimating local economic impacts of recreation and tourism. In S. Reiling (Ed.), *Measuring tourism impacts at the community level* (Maine Agricultural Experiment Station Miscellaneous Report No. 374). Orono, ME.

Stynes, D. J., & Propst, D. B. (1996). *MI-REC manual version 3.0.* East Lansing, MI: Department of Park, Recreation and Tourism Resources, Michigan State University.

Stynes, D., Propst, D., Chang, W., & Sun, Y. (2000). *Estimating national park visitor spending and economic impacts; the MGM2 model.* East Lansing, MI: Michigan State University.

The Outspan Group. (2011). *Economic impact of parks Canada.* Amherst Island: The Outspan Group Inc.

Thapa, B. (2013). Introduction to the special issue. Special issue on managing protected areas: Global perspectives, volume I. *Journal of Park and Recreation Administration, 31*(2), 1–6.

Turpie, J., Barnes, J., Lange, G., & Martin, R. (2010). *The economic value of Namibia's protected areas system: A case for increased investment.* Namibia: Ministry of Environment and Tourism.

U.S. Dept. of Commerce, Bureau of Economic Analysis. (1992). *Regional multipliers: A user handbook for regional input-output modeling system (RIMS II)* (2nd ed.). Washington, DC:; U.S. Gov't Printing Office.

U.S. Dept. of Interior, National Park Service. (1990). *The money generation model.* Denver, CO: Office of Social Science, Socio-Economic Studies Division.

U.S. Travel Data Center. (1997). *Impact of travel on state economies, 1995.* Washington, DC: U.S. Travel Data Center.

Viveiros de Castro, E., Souza, T. V. S. B., & Thapa, B. (2015). Determinants of tourism attractiveness in national parks of Brazil. WCPA (2015). PARKS. *The International Journal of Protected Areas and Conservation, 21.2*(51–62).

World Travel and Tourism Council. (1996). *State of Hawaii travel tourism; A new economic perspective.* London: Author.

Khanyayo village and Mkhambathi Nature Reserve, South Africa: a pragmatic qualitative investigation into attitudes towards a protected area

Dorothy Queiros and Kevin Mearns

ABSTRACT

Protected areas are often surrounded by impoverished communities. Biodiversity must be conserved while enabling community members' livelihoods. Benefit-sharing has been promoted as a means of fostering positive relationships between protected areas and local people, but views differ regarding which benefits have positive results; and which are most valued by communities themselves. This research investigates the attitudes of Khanyayo village towards Mkhambathi Nature Reserve on the Wild Coast of South Africa. A pragmatic, multi-method comparative qualitative approach was followed. Participants were local community members and key staff at Mkhambathi, who work with the community. Individual interviews, focus group interviews and an adapted nominal grouping technique were used, which aided triangulation. Results indicated a range of significant benefits and losses. Tangibles included access to natural resources, employment, infrastructure and training. More intangibles emerged though, such as enjoyment of reserve, involvement in the reserve, communication and environmental education. Losses predominantly related to prohibited or restricted access to natural resources. The comparison between the two participant groups yielded interesting insights. Knowing which benefits are most important to the community and aligning the perceptions of the people and the park staff could improve the relationship as well as enhance successful benefit-sharing.

Introduction

The future success of conservation efforts largely depends on the attitudes and behaviour of communities living in and around protected areas (Snyman, 2014). The natural environment globally faces a crisis in terms of biodiversity loss, making conservation of protected areas essential (Abrams et al., 2009). At the same time, most of South Africa's protected areas are surrounded by impoverished communities (Davies, Fazey, Cresswell, & Pettorelli, 2014). Biodiversity must be conserved while enabling the livelihoods of local communities (Davies et al., 2014). This is challenging and complex, as communities living in or around protected areas often depend on natural resources for survival and income (St John, Edwards-Jones, & Jones, 2010).

Previous research identifies gaps in scholarly knowledge on what encourages communities to conserve and have a positive attitude towards protected areas (Berkes, 2004; Coria & Calfucura, 2012;

Imran, Alam, & Beaumont, 2014; Kiss, 2004). Benefit-sharing appears to play a role, but the influence of individual benefits is unclear. To contribute to current understanding of the above, this research aimed to determine attitudes related to benefits and costs, focusing on the case of the Khanyayo village[1] and Mkhambathi[2] Nature Reserve on the Wild Coast of South Africa.

This research follows a novel methodological approach, in accordance with what Gubrium and Holstein (2014, p. 35) term "explanatory excitement", advocating for a move beyond methodological strictures to new ways of seeing. They contend that what matters is to provide understanding. A multi-method comparative qualitative approach was used to effectively conduct research in a rural setting. The study borrows from interpretivism and pragmatism, which gives researchers freedom to choose methods that best draw out the data required to answer the research question. This does not allow though for procedural recklessness – rigorous, systematically designed and applied methods are important.

With attitude studies in this field primarily being quantitative and focusing on communities only, this research engages two different constituencies – the local community and the reserve – to learn from both the people and the park. Since each has very different contexts, insights are provided from both sides that increase our understanding, provide a better triangulated response to these dual realities and yield some interesting lessons.

Context based on previous research is provided first. Although this study focuses on attitudes, these are often linked to actual behaviour, hence these linkages are acknowledged, and also emerged in this research. This is followed by the case background, research design and methodology, results and discussion, and a conclusion with recommendations.

Improving relations between communities and protected areas

Africa's population is likely to double to 2.4 billion by 2050 (Baker, 2015). By 2040, it is estimated that half the world's population under the age of 24 will be in Africa. This group constitutes 60% of the continents jobless inhabitants (African Development Bank, 2011, cited in Baker, 2015). This "demographic time bomb" could threaten Africa's protected areas.

The dual crisis of poverty and biodiversity loss is of global concern, leading to attempts to decrease poverty through communities becoming involved in and supporting biodiversity conservation (Abrams et al., 2009; Davies et al., 2014; Nyaupane & Poudel, 2011). However, McShane et al. (2011) acknowledge that such win–win solutions are hard to achieve, and advocate making trade-offs and accepting the limitations of achieving both these goals. Torquebiau and Taylor (2009, p. 2543) concur, stating that "people-based conservation is the only realistic alternative for the long-term management of natural resources, even if it sometimes fails".

To reduce opposition and attract local support, benefit-sharing approaches have been widely adopted, based on the idea that benefits motivate local people to change attitudes and move towards pro-conservation behaviour (Gadd, 2005), while lack of benefits or costs/losses cause negative feelings and destructive environmental behaviour (Burgoyne & Mearns, 2017). Benefit-sharing (or benefit-based approaches) in the context of this paper occur when the protected area shares benefits with the local community for the purposes of improving their social and economic environment, and to foster a good relationship. Benefits can be tangible and intangible and include revenue sharing, access to natural resources, infrastructure development, sharing of information, collaboration, education and training initiatives, etc. The initiatives can come directly from the protected area or offered in collaboration with other organisations and government bodies.

Mehta and Heinen (2001) cite growing evidence that local peoples' support for protected areas depends predominantly on their perceptions of costs and benefits from living around these areas. As an example, Kideghesho, Røskaft, and Kaltenborn (2007), in their study in Tanzania, found that communities experiencing more wildlife induced costs were less likely to support conservation; while those receiving more wildlife-related benefits were more likely to support it.

Some scholars support the linkage progression that benefits received, result in positive attitudes towards the environment (Liu et al., 2014; Reimer & Walter, 2013). This in turn can lead to pro-conservation behaviour (Shibia, 2010; Tran & Walter, 2014), but is difficult to measure (Snyman, 2012). Others have proven that benefits result in positive attitudes, but not conclusively in positive pro-conservation actions (Kiss, 2004; Snyman, 2014). Further studies found that benefits received do not necessarily cause positive attitudes (Waylen, Fischer, Mcgowan, Thirgood, & Milner-Gulland, 2010), let alone positive behaviour.

The types of benefits and costs which influence pro-conservation attitudes are briefly discussed below and later reflected on in the context of the findings.

Tangible benefits appear to be important to provide communities with vested interest in conserving the land and a reduced incentive to practise destructive land-use forms (Langholz, 1999; Mbaiwa, 2005). Tangible economic benefits emerging from literature include employment (Shibia, 2010); stable livelihoods (Stoll-Kleeman, 2005); and multiple livelihood strategies (Gurung & Seeland, 2011). "Stable livelihoods around a protected area are the best pre-conditions for local acceptance of use restrictions inside the park" (Stoll-Kleeman, 2005). Alternative sources of income can take diverse forms, from better access to tourism-related jobs to better cultivation techniques (Stoll-Kleeman, 2005). Gurung and Seeland (2011) and Snyman (2014) suggest that multiple livelihood strategies are important to avoid exclusive dependence on natural resources for survival. Small-scale and locally owned ecotourism is one means of creating diverse livelihood portfolios (Nyaupane & Poudel, 2011). The ideal solution to long-term conservation success lies in improving local peoples' living standards so they are no longer entirely dependent on proximate natural resources for survival (Snyman, 2014). Benefits will not deter people from illegal activities if they cannot meet basic needs. Conversely, Waylen et al. (2010) identified natural resource use as a significant tangible non-economic benefit, determining that community-based tourism success was not predicted by economic and practical benefits, but by use and control of natural resources, resulting in greater support and concern for conservation. In their paper on attitudes towards the Kgalagadi Transfrontier Park, Botswana, Moswete, Tapa, and Child's (2012) conclude that power over the management of natural resources needs to be shared, and that top-down approaches disempower local people.

Intangible benefits (social and individual) are also significant. Campbell, Haalboom, and Trow (2007) found that although economic benefits were important, community benefits such as unity and community learning are often more highly valued. Stem et al. (2003) reported that indirect tourism benefits and education levels show stronger associations with conservation behaviour and perspectives. Gadd (2005) and Stem et al. (2003) determined that income generation alone is insufficient, with indirect tourism benefits, particularly ideas exchange, training and equity in benefit distribution and empowerment being more likely to cause pro-conservation behaviours. Berkes (2004) concurs, stating that rural communities in developing countries seldom equate benefits with simple monetary rewards and that various types of social and political benefits are also important. "Social improvement benefits" such as clinics (Burgoyne & Mearns, 2017) and building on local aesthetic values and traditional beliefs (Gadd, 2005) are also mentioned. Matarrita-Cascante, Brennan, and Luloff (2010) refer to "community agency" which is created by relationships, interaction and communication. These increase the adaptive capacity of a community, impacting on attitudes and behaviour:

> ... community agency resulted in the dissemination of pro-environmental attitudes and actions in the community.... Sustainable goals were often communicated and discussed in ...gatherings. With time, residents exhibited a shifting and growing concern towards protection of the natural environment, [resulting] in ...protection of nature, which in turn sustained ...tourism. (Matarrita-Cascante et al., 2010, p. 749)

Kideghesho et al. (2007) concur, highlighting regular contacts between local communities and project staff as a means of achieving positive attitudes. In the attitude study by Moswete et al. (2012), locals lamented the limited communication regarding what was going on in the park, which suggests that its absence could negatively affect attitudes.

Information and education on tourism, resource use and environmental impacts also appear to be intangible benefits (Imran et al., 2014; Kideghesho et al., 2007) and can change attitudes (Waylen et al., 2010) and encourage long-term conservation by local people (Stem et al., 2003). Finally, Matarrita-Cascante et al. (2010) claim that the presence of tourism can play a role in encouraging environmental consciousness and protection because of the realisation that nature brings tourists.

In terms of *losses/costs*, human-wildlife conflict (crop and livestock losses, threats to human safety and/or property damage) recur in literature, appearing to negatively influence attitudes (Karanth & Nepal, 2012; Nyaupane & Poudel, 2011; Snyman, 2012).

Just as access to natural resources is seen as a benefit, so restricted/denied access is viewed as a loss, impacting at household level or resulting in lost economic opportunities (Gurung & Seeland, 2011; Stronza & Gordillo, 2008). In Karanth and Nepal's (2012) study in India and Nepal, as well as Kepe's Mkhambathi studies (2004, 2007), for example, restricted access to non-timber forest products, grazing and fuelwood, was found to cause contention. Karanth and Nepal found that resource dependency was a main predictor of attitudes, with locals who were not dependant on protected area resources being more favourable towards the protected area. Bennett and Dearden (2014) report that National Marine Parks in Thailand had significant impact on livelihood strategies, with participants not being able to fish or harvest for subsistence purposes or to make a living. Participants reported that this left locals with no other livelihood options but to "steal" from the sea, resulting in negative perceptions towards these parks.

Other costs highlighted by Stronza and Gordillo (2008) include leaving the family or farm and losing connections with community (when lodges are far from villages); problems with co-workers; eating unfamiliar foods; working long hours; and erosion of traditional relations of cooperation and reciprocity. Other socio-cultural costs/losses include drugs and cultural distortions (de Los Angeles Somarriba-Chang & Gunnarsdotter, 2012); children not wanting to dress traditionally; and tourism encouraging greed amongst neighbours (Gurung & Seeland, 2011). The increased prices of goods/services are another negative (Lee, 2013); as well as conflict between visitors and locals (Lee, 2013), park management and locals (Gurung & Seeland, 2011), and between locals (Simpson, 2008). Environmental costs in the form of damage from tourism (Gurung & Seeland, 2011) and the cost of restrictions on time due to tourism involvement (de Los Angeles Somarriba-Chang & Gunnarsdotter, 2012) also emerge.

In this research, attitudes (also referred to as perceptions) are defined as "respondents' particular feelings and perceptions towards the stated questions relating to tourism and/or conservation" (Snyman, 2014, p. 2 – adapted from Allport, 1996 and McDougall & Munro, 1987). Pro-conservation behaviour, though not the focus of this article, is also defined because of the attitude-behaviour link. It refers to behaviour that harms the environment as little as possible and could even benefit it (Steg & Vlek, 2009).

Background to case study

Mkhambathi Nature Reserve (MNR) is a 7720 ha reserve in the Eastern Cape of South Africa, on the Wild Coast. It consists of coastline (Indian Ocean); open grassland with patches of indigenous and swamp forest; several grazing herbivores (e.g. Eland and Blesbok); and large rock pools and spectacular waterfalls, one of which falls into the ocean (Queiros, 2000). The coastline is within the Pondoland Marine Protected Area, and the reserve forms part of the Maputaland-Pondoland-Albany biodiversity hotspot, which is internationally recognised for its high levels of endemism and species richness (Blore, Cundill, & Mkhulisi, 2013).

To the west of the reserve lies 11,000 ha of state land. A government parastatal known as TRACOR (Transkei Agricultural Corporation) used this land for sugarcane and cattle farming. Apart from a community forestry project, it is currently largely unutilised.

Beyond TRACOR lies the predominantly rural communal settlement, comprising the Khanyayo Administrative Area (Fihlani, 2015). It constitutes seven villages inhabited by the Khanyayo people,

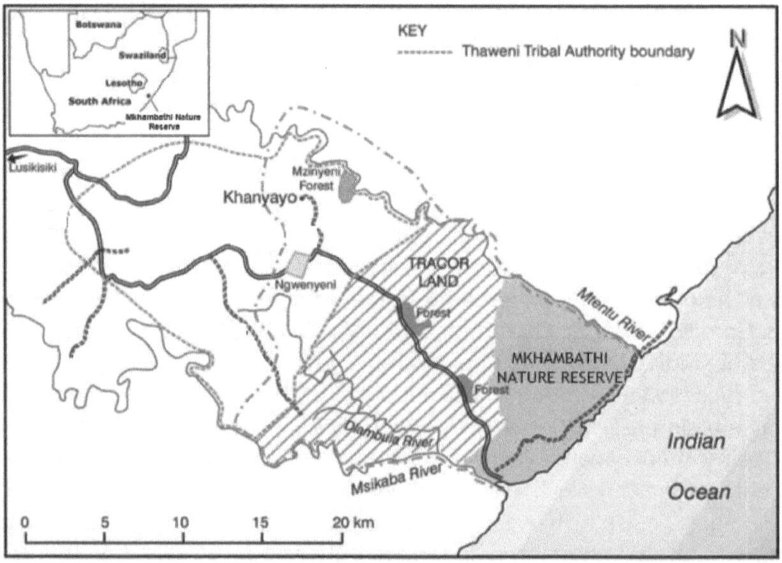

Figure 1. Map showing the location of Mkhambathi Nature Reserve and Khanyayo village (adapted from Kepe, 2008. Used with permission).

who speak isiMpondo, a Xhosa dialect (Kepe, 2007). The villages are Khanyayo, Mtshayelo, Rhamzi, Cele, Thahle, Ngquza and Vlei (Zeka, 2011) and form part of the Mkhambathi Land Trust and the Thaweni Tribal Authority (TTA) (Kepe, 2004). The closest village to the reserve is Khanyayo village (approximately 17 kms away), in which the research was conducted. According to the current reserve manager, the population of Khanyayo village is over 3000 inhabitants. Figure 1 shows the positioning of Khanyayo village, TRACOR land and MNR.

Forty-six per cent of the population within this administrative area is 15 years or younger, while 48% are between 15 and 64 years old. The area has low educational levels and poor literacy rates, with only 2.4% having matric (final school year), and 1.4% post-matric qualifications. Functional literacy is estimated at 48%, and unemployment at 78%. Over 97% of households live on less than R800 per month (Fihlani, 2015), generating livelihoods from a combination of livestock, arable farming and collecting natural resources. Remittances and pensions also supplement livelihoods (Kepe, 2007).

Regarding the history of the community and the reserve, in 1899 the paramount chief (Sigcawu) of the Mpondo kingdom, under whose jurisdiction the area fell, agreed with government that a portion of land could be used as a leper colony. In 1920, the Khanyayo people were removed from the colony, marking the beginning of a long conflict (Kepe, 2004). In 1976, the land became a privately run nature reserve and in 1982, the reserve became government-run. A formal land claim from Khanyayo village was gazetted in 1998, but disputed by the other villages within the TTA, who lodged a counter claim. Conservation was severely affected during this time, as the TTA forcefully occupied reserve buildings, and increased illegal use of flora and fauna within MNR. Herds of wild herbivores were driven over cliffs and vegetation burnt (Kepe, 2004). In 2001, the Land Claims Commission identified 326 households as descendants of those originally removed from the reserve, who would be financially compensated. It also ruled that the approximately 6000 households[3] (Kepe, 2012) of the seven villages within TTA would together share joint management and have joint land rights over Mkhambathi, receiving financial and other benefits. In October 2004, the formal handover of reserve and TRACOR land to the Mkhambathi Land Trust took place (Kepe, 2012).

Today MNR is managed by Eastern Cape Parks and Tourism Agency (ECPTA) and the Mkhambathi Board of Trustees, which consists of two members from each of the seven villages. Key reserve staff attend community meetings where information is conveyed and community concerns heard. The reserve manager is well known and respected in the community.

MNR runs several benefit-sharing initiatives, some of which are in partnership with other bodies. These include, among others, access to grass for thatching; access to enjoy the reserve as visitors; the Environmental Schools Competition and environmental education in schools; sports programmes; workshops which train community members in various modules such as communication and administration skills, computer training and project management, and environmental education; job opportunities (permanent jobs and temporary contracts, for example as part of "Working for Water, 'Working on Fires" and "Coast Care Project"); revenue sharing of 15% of income accruing to reserve; fire awareness training; and skills development training in hospitality, field ranging and infrastructure maintenance.

MNR offers limited self-catering options for tourists (two 6 sleepers and one 10 sleeper). A portion of the reserve is currently under development with a private company.

Research design and methodology

The question guiding this research was "What influences the attitudes of the Khanyayo community towards Mkhambathi Nature Reserve (MNR)?".

The research design borrows from interpretivism and pragmatism (Creswell, 2014), and follows a comparative multi-method qualitative research design. Quantitative research on attitudes is common, yet, qualitative approaches hold promise since they provide the means to understand, in their own words, the feelings, values, perceptions, opinions and lived experiences of participants that underlie and influence attitudes and behaviour towards the natural environment (Patton, 2005; Roulston, 2014). The interpretivist paradigm is growing in acceptance and needs to be further developed in the South African context.

Participants

Two different groups of participants (termed "constituencies") took part in this research: Constituency 1 (the local community living in Khanyayo village) and Constituency 2 (ECPTA representatives working inside the reserve who had worked with the local community). These are referred to as C1 and C2, respectively.

Data gathering with *C1 participants* involved half a day spent with participants at a local school in Khanyayo village, in December 2015. This village was chosen, because it is the closest to the nature reserve (hence more opportunity of interaction between this community and MNR than with others[4]); and forms part of the Mkhambathi Land Trust.

When permission for the research was sought from the chief, he was provided with criteria for the C1 participants (aged 18 or over, a mix of age and gender, and literate – in order to participate in nominal grouping technique (NGT)). Through the open invitation of the chief to community members who fitted these criteria, various sectors of the community were present. This process effectively led to a non-probability purposive sample. Twenty seven participants responded to the invitation, with the final group comprising the chief, some community leaders and others, with more females than males present. In rural research in South Africa, it is protocol to work through the chief. The reserve manager therefore approached the chief first, on behalf of the primary researcher. Bias that could emerge from the chief being part of the sampling was mitigated due to him issuing an open invitation. The primary researcher was satisfied that the final group comprised a healthy mix of ages and positions in society. Having more females than males indicated a lack of purposeful influence by the chief, as community leaders in this area are usually men.

This initial group then divided themselves into two groups for the focus group interviews (FGI). While one group was interviewed, the other withdrew to another classroom to draw a map[5] showing

Table 1. Participants and data collection.

	C1			C2
	Focus group 1	Focus group 2	Focus groups 1 and 2	C2
Number of participants	15	12	19*	5
Gender	11 F, 4 M	10 F, 2 M	13 F, 6 M	1 F, 4 M
Data collection method	Focus group interview	Focus group interview	• Nominal grouping technique • Final question via combined focus group interview	Individual semi-structured Interview

*Eight participants left before the NGT session.

their village and MNR. The groups then swapped. The mapping group could not hear the responses of the interview group. NGT was undertaken later with both groups together.

C2 participants were representatives of ECPTA working at Mkhambathi, and who worked closely with the local community. In a small reserve, selection needed to be pragmatic and was via the researcher's requests and the reserve manager's advice. This also constituted non-probability purposive sampling. Five participants holding key roles at MNR were interviewed via semi-structured individual interviews (II), at the reserve itself. In total, Mkhambathi has 27 employees. Due to the low number of ECPTA employees working at a small reserve like MNR who *also* work with the local community, it was impossible to replicate the sampling strategy for C1 (a group of 20+ participants). Conversely, it was undesirable to have five interviews with C1 (replicating C2's methodology), as FGIs are more appropriate in the African context, where discussion and joint generation of ideas is common. Furthermore, the C1 FGIs and NGT were deemed the best ways to gather multiple perspectives from several C1 participants simultaneously; while within C2, IIs provided more depth, the opportunity for the researcher to engage with a topic for longer, and valuable historical background.

Limitations of the study are that it was not a village-wide study and that, by following protocol, the chief played a role in participant selection, which could lead to bias. An assumption was that all participants answered the questions truthfully.

Table 1 summarises the final sample and data collection methods used.

Multiple methods

The different methods used have been mentioned above but are briefly motivated below. Cohen, Manion, and Morrison (2011) advocate eclecticism and fitness for purposes in qualitative data collection – to explore phenomena more fully, provide greater depth and to triangulate by seeking different data on the same phenomena. They further contend that triangulation via multiple methods is less vulnerable than a single-method approach. Grbich (2007) argues for innovative mixes of data sources that allow one to view phenomena from different perspectives. A multiple method study by Darbyshire et al. (2005) concluded that the approach was valuable in terms of confirming data as well as providing complementary insights and understanding that would be difficult to achieve via a single method.

The method used for each question is shown in Table 2, and its choice depended on the style of answers likely to be elicited. For *C1 participants*, some questions were asked via FGI, and the responses recorded. The primary researcher took on the role of facilitator, with assistance from the translator where necessary. FGI's were ideal when discussion and insight was required, and where feelings were likely to emerge. They elicited responses from a variety of participants, encouraging interactive discussion, which revealed complementary and contrasting opinions. The use of two FGIs provided more opportunity for patterns to emerge.

An adapted form of NGT was also used with C1. NGT is a consensus method helpful in synthesising individual opinions, without the limitations of group interaction where certain individuals may dominate (Van Tejilingen, Pitchforth, Bishop, & Russell, 2006). In original NGT, individuals generate

Table 2. Questions in research instrument and associated data gathering method.

Q	Question	Asked to:	Data gathering method
(1)	What do you know about this nature reserve? What is inside this nature reserve? What can you do in there?	C1	FGI
(2)	Tell me about the relationship between you and the nature reserve. How do you feel about living near the nature reserve?	C1	FGI
	Tell me about your perceptions of the relationship between the local community and the nature reserve.	C2	II
(3)	How has the nature reserve changed the way you live (positive and negative)? How have things changed?	C1	FGI
	How do you think the nature reserve has changed the way the local community lives (positive and negative)? How have things changed?	C2	II
(4)	Some people like this nature reserve and the animals. Some people think there are better ways to use this land. What would make you more positive towards the nature reserve being here over the next 100 years, that is, down to the time of your great grandchildren?	C1	FGI
	What do you think would make the local community more positive towards the nature reserve being conserved in the future?	C2	II
(5)	What do your friends and family think about this nature reserve?	C1	FGI
(6)	Who of you have been into the reserve? How many times a year? What do you go in for? What did you think of your experience?	C1	FGI
(7)	Do you have any responsibilities for this nature reserve? If you do, how do you feel about these?	C1	FGI
	Do the local community have any responsibilities for/towards this nature reserve? If they do, how do you think they feel about these?	C2	II
(8)	What are the benefits of having this nature reserve near to your home? Which of those benefits are most important to you? Which are least important?	C1	NGT
	What are the benefits to the local community of living near this nature reserve? Which of those benefits do you think are most important to them? Which are least important?	C2	II
(9)	What are the losses/costs of having this nature reserve near to your home? Which of those costs impact the most on you? Which ones impact the least?	C1	NGT
	What are the losses/costs to the local community of living near this nature reserve? Which of those costs do you think impact the most on them? Which ones impact the least?	C2	II
(10)	For you, living near this nature reserve, what is your ideal future for your community? What is your dream situation?	C1	FGI
	For this local community, living near this nature reserve, do you have any ideas on what could be an ideal future for them?	C2	II

Key: C1 = Constituency 1 (Local community).
C2 = Constituency 2 (ECPTA representatives).
FGI = Focus group interview.
II = Individual interview.
NGT = Nominal grouping technique.
Q = Question.

ideas which are pooled, discussed, organised and finally voted on (Chapple & Murphy, 1996). The primary researcher found little evidence of NGT being used in rural settings, but experienced that a simplified version was highly effective in gathering data. Participants were provided with several sticky notes, with females and males receiving different colours. They were asked to generate as many answers as they wished, as individuals with no group discussion. In this approach, answers could be a single word or sentence, in mother tongue or English, and were usually short responses. Participants then stuck the notes on a large paper and together organised them into categories, naming each category. The notes in each category were tallied and participants were asked whether they agreed with the order of priority (according to the tally). If not, this was voted on. NGT was useful for questions where the researchers wanted to generate lists/multiple answers, e.g. benefits received, and provided a means to rank benefits/losses via categorisation and voting. Younger females, who were less participative in FGIs were very active here. This could indicate that simplified NGT is a powerful data collection method to hear the voices of the less outspoken in rural research.

For *C2 participants*, data gathering was via semi-structured IIs, ranging from 45 to 66 min in length. These were recorded. In general, the quotes emerging from these are longer and provide additional depth and examples.

The final question (regarding an ideal future) was asked to C2 in the same form mentioned above. For C1, both focus groups joined and each participant was asked to articulate their personal dream for an ideal future.

Research instrument

Existing literature was consulted before developing the research instrument, which investigates the research question from different angles. The instrument was reviewed by subject specialists and piloted in an earlier study at Dinokeng Game Reserve (Gauteng Province, South Africa).

The questions asked are provided in Table 2. C1 and C2 participants were asked questions on the same topic. C1 were asked to reflect on their personal opinions, while C2, who all had experience working with this community, were asked what they perceive the community's responses to these issues to be. Since C2 participants worked with the reserve *and* community, they were in an ideal position to provide opinions on how they think C1 perceive these matters. C1 and C2 are both crucial role-players in the people-parks relationship, and the pragmatic stance determined that the research would be incomplete without the parks perspective.

Some questions, not relevant to C2, were only asked to C1.

A local translator was used to reduce the risk of the researcher creating bias and impacting trustworthiness, and to set participants at ease (Nault & Stapleton, 2011; Snyman, 2014). Anonymity was assured. The first questions were easy, settling participants in; while NGT encouraged participation from the more reserved. The use of different tools and constituencies served to reduce research bias and verify the data, thus improving trustworthiness.

Ethical clearance was awarded by the Research Ethics Review Committee of the College of Agriculture and Environmental Sciences at the University of South Africa (UNISA). Written permission was obtained from the chief, Reserve Manager and ECPTA. Participants signed informed consent and could leave at any time.

Data analysis

A computer assisted data analysis system (CAQDAS) package, ATLAS.ti, (Version 7.0) was used to manage data and organise the analysis (Gibbs, 2014). ATLAS.ti provides tools that enable, among other things, organisation of large volumes of different data types, transcribed text to be linked back to its exact spot within original recording, coding, memo and comment writing, various search and data organisation options, visualisation of data, and the choice of various analytical tools. Within ATLAS.ti the recordings of the C1 FGIs and C2 IIs were transcribed. NGT data were captured, with each note forming a single document (93 documents in total) within ATLAS.ti.

Coding of the above documents was open and inductive. For each question, a code set was developed and coding rules articulated to ensure consistency across documents. The coding system was refined as coding progressed, culminating in a universal coding frame (containing code sub-sets for individual questions) that was used to code all data across all methods. Hence, whether data had been collected via a C1 FGI, note generated during NGT, or from the combined FGI used for Q10; or a C2 semi-structured II, all were coded using the same coding frame.

Data analysis followed by determining similarities, differences, relationships and patterns; and developing a small set of themes that covered these (Miles, Huberman, & Saldaña, 2014). Cohen et al. (2011, p. 537) assert that "There is no one single or correct way to analyse and present qualitative data; how one does it should abide by fitness for purpose". With multiple methods an integrated approach is required, involving a greater degree of synthesis (Coles, Duval, & Shaw, 2013; Creswell, 2014). Researchers need to organise it into themes that cut across all data sources (Creswell, 2014). Cohen et al. (2011) argues that "cutting across" different sources is necessary in order not to lose connections between data. The reporting of multiple perspectives develops a complex holistic account (Creswell, 2014). Palmberger and Gingrich (2014) refer to comparison in qualitative studies as a means to increase understanding across a plurality of approaches. Barbour (2014) in fact, views neglecting the potential to compare, as a weakness in qualitative analysis.

Hence, this research compared C1 and C2 to provide a more comprehensive idea of the issues at hand and to triangulate the data. Without doing this, insights into the relationship between Khanyayo villagers and MNR would have been lost.

At all times, the quotes formed the basis. Quotation frequencies were used when these added additional insight, referred to as "quantifying your qualitative data" (Saunders, Lewis, & Thornhill, 2009). Miles et al. (2014) and Cohen et al. (2011) report an established tradition of dealing quantitatively with qualitative data, stating that counting is helpful to quickly see the content of a complex batch of data; to keep the researcher analytically honest, protect against bias; and to ensure that a theme is indeed important or recurring. Barbour (2014) even argues that qualitative researchers should show evidence of counting as a form of reassurance.

Hence this research has used a combination of quotes and counting frequencies where valid. At all times, the original quotes took precedence, and were studied first to determine if quantification was valid. For example, if a quote indicates that no benefits are received, then it is not counted as a quote under "Benefits", as this misleads the reader into thinking that "X amount of benefits were mentioned". For each question, after coding, the data were examined graphically, comparing C1 and C2 in terms of what they had said (quote content) and how often this emerged (counting frequencies), searching for the overriding prevalent themes that cut across the data and across the individual questions asked.

Results and discussion

Results are not discussed per question, but according to key themes emerging, which are presented in Table 3. The table also provides the codes used to code the quotations, and from which the themes were gleaned; the question number (Q), corresponding with the question numbers in Table 2; and data (quotation) tallies for C1 and C2. For example, the code "Visiting the reserve: beach" developed for Question 8 was used to code 12 quotes from C1, and two from C2. A similar code also emerged inductively in Question 6's coding set. Ultimately, all references to visiting MNR for leisure were combined into a theme, "Enjoyment of reserve".

Results are interpreted below. Quotations, the data in qualitative studies, are used to motivate the key themes. Findings are also compared with previous studies.

Access to natural resources

Access to natural resources emerged from C1 and C2 as the most significant benefit (in Q8) and as the most mentioned positive change (Q3). Quotes reveal that communities view access to natural resources as a positive gesture from MNR. Waylen et al. (2010) found that allowing communities to use protected areas was more important than economic and practical benefits, resulting in greater support and concern for conservation.

Being able to collect grass within the reserve (for thatching huts) was the resource most mentioned by C1, and corroborated by C2. Regarding C1, for males no single benefit stood out from the rest, while for females, collection of thatch grass was one of the most significant. Since this question was answered by C1 using NGT notes, their responses are mostly short, naming the benefit. The depth from the C2 quotes is therefore mainly used to illustrate this theme.

> Because the one that I will say [is most important to the communities] ...is the grass harvesting [when the season is open], you will see lots of people coming in for that. Lots of old people, old ladies, they come and stay here for the whole period.... There's a house for them to come and bring their mattresses and ...stay there. Sometimes they even ask for the period to be extended. So I will say it's one of the things that they want, they know that they're going to benefit. So from that they go and build their houses free, because zinc is expensive.... .Grass is much better. (C2)

Table 3. Quotation frequencies for key themes.

Theme	Code	Q	Quotation frequency C1	C2
Access to natural resources	Lack of access to natural resources	1	5	N/A
	Access to natural resources	3	2	6
	Lack of access to natural resources	3	0	5
	Access to natural resources	8	17	11
	Lack of access to natural resources	9	38	5
Enjoyment of reserve	Beach/swimming	6	5	N/A
	Animals	6	3	N/A
	Visiting reserve: beach	8	12	2
	Visiting reserve: seeing animals	8	2	0
	Visiting reserve: waterfalls	8	1	0
Intrinsic appreciation of reserve	Appreciate reserve/resources	2	4	2
Community projects	Community projects	3	0	4
	Community projects	4	0	6
	Community projects and financial aid	10	0	2
Employment	Employment	3	1	6
	Less jobs than before	3	1	0
	Employment	4	5	4
	Benefit from jobs	8	7	8
	More development and employment	10	5	1
	Facilitate entrepreneurship	10	0	1
Training	Education/training	3	0	4
	Past training: Locals protect reserve	7	0	2
	Learning/training: New skills	8	1	2
Development and infrastructure	Facilities and infrastructure	3	1	2
	Development/infrastructure	4	10	0
	Positive: But want development	5	1	0
	More development and employment	10	5	1
Communication and information sharing	Lack of knowledge/information/access	2	0	1
	Frustration from lack of action	3	1	1
	Communication	4	1	4
	Negative: Want information	5	3	N/A
	More communication with reserve	10	2	0
Involvement and responsibility	Involvement/interaction	4	2	1
	Locals protect reserve	7	0	6
	Ideas from locals to encourage protection	7	4	0
	Locals protected reserve in the past	7	2	0
Environmental education and awareness	Information/education	4	1	5
	Learning/training: About environmental awareness/education	8	2	3
	Environmental education	10	1	1

Key: C1 = Constituency 1.
C2 = Constituency 2.
Q = Question.

> The thatching ...it's a big benefit, because they used to get this thatch grass around their villages. But because of their land management in terms of fires, they lost that. So that's why they can only get it now here at the reserve. Because it is still protected [and] managed properly. You can't even get it just outside the gate. So you can imagine if it was not a nature reserve ... .that specie [would] be gone. (C2)

Collecting firewood – in TRACOR land and when MNR cuts exotics – also emerged. Although these are not resources for harvesting, access to ancestral graveyards and the use of the sea for rituals were also mentioned, where complex requirements are met by MNR (e.g. allowing the amagqirha, the Xhosa traditional healers, to access the ocean at midnight):

> ... there [are only] a few graveyards that people ...still have. But they still have access to those. That one, we don't even debate. The [amagqirhas] [can come in] anytime. They don't even need to pay to come and do their rituals, ...because we want to bring it back as it was ... .They will have a person to heal and then they have to come to the reserve [at] midnight. ...we will just try to arrange it because that is the time that has been [determined], and we can't change that. (C2)

Just as resource access emerged as an important benefit (where MNR management facilitate controlled access to certain resources), it also manifests as a significant loss (where resource access is restricted or prohibited) in Q9 with the majority of mentions coming from C1 (C1: 38; C2:5), and as a negative change in Q3. In Q1 (knowledge of the reserve), lack of access to resources also surfaces. Not being able to collect medicinal plants emerges most. For C1 males, this was also their most prominent loss. Examples of C1's NGT notes refer to "Some people used to dig some medicine there (traditional ones) but now it's not allowed"; and "We can't get the herbs now, since there's Mkhambathi". Longer quotes from C2 elaborate:

> ... some of the medicine they want is in the reserve - but there's also medicine in the forests outside the reserve. ...There's something that they always cut and take to Durban and sell. Because the problem is selling. They collect it ...and sell it. It's not that they take something small and use it for healing, maybe. But they make a business out if it, that's why it's a problem. The area which is outside the reserve is big enough for them to go and collect their stuff.

> ... what you'll normally hear from the people that are more traditional, they know the muti[6] [and feel that] the nature reserve is preventing the harvesting of those type of medicines. ...people from local villages, they always say, "no, we want those traditional medicines and plants inside and trees, but the park is preventing us from that, but it's our land and everything". But awareness is being done with them why we're preventing [this], but ...it's what ...local people are saying.

The inability to hunt emerges as the second most significant loss, e.g.: "We can't hunt now because there's Mkhambathi" (C1). For C1 females, this was the loss that emerged above others. For C1, inability to hunt was followed by restricted fishing. Four quotes mention that one cannot fish, whereas three referred to restricted fishing [e.g. "Lots of restrictions like no fishing without permits" (C1)]. The latter is the reality, as explained below by a C2 participant:

> There is this issue of access in terms of when they want to come and fish. Out of the restrictions that [exist], you can fish in some places, not in other places, [there are] bag limits now which they didn't used to have.

The loss of land and desire to use the land for other purposes was also mentioned, for example:

> See, some of community they don't want the reserve, because they compare this reserve to the other side of KwaZulu Natal, they said, "no, there's houses next to the sea and they can build shacks". So this is what they want to do, but we tell them that it's not allowed. And they also want their cattle to come and eat grass inside the reserve. (C2)

Finally, being unable to access places of spiritual importance emerged as a loss from C1, who mentioned being unable to visit ancestor's graves, e.g. "Cannot visit our ancestors". However, as mentioned by a C2 participant above, this is permissible by prior arrangement with the reserve.

In terms of losses, it appears that access to natural resources for these rural communities is very important. This is in line with other studies such as Gurung and Seeland (2011) and Karanth and Nepal (2012). It is noteworthy that C1 are more vocal than C2 on the matter.

Enjoyment of reserve

In Q6 (experiences of reserve), C1 participants are highly positive about their enjoyment of MNR, particularly the beach, followed by animals. All C1 participants (except two) had been into the reserve for the beach. When naming benefits received (Q8), beach access was mentioned 12 times by C1, and twice by C2. For the females in C1, along with collecting thatch grass, visiting the beach was the other most significant benefit. When voting on the benefit most important to them (since collecting thatch grass and beach access both had 12 mentions), C1 unanimously placed beach access in first place. In the IIs, C2 respondents were asked in what order they would rank the benefits they had outlined. Those who mentioned the beach ranked it as third and fifth respectively.

References from C1 to visiting waterfalls and seeing animals also emerged under Q8. Again this was not mentioned by C2.

Intrinsic appreciation of reserve

It is significant that overall, C1 articulated more benefits than losses (via NGT). In Q2, an appreciation of the reserve surfaced.

> There's not an issue of saying they don't want this reserve ...some people would say so, but I would say the majority say it is good to have it. (C2)

> I would say we feel so special that the nature reserve is in our community; and tourists have to come all the way from ...the inland areas to spend quality time with their families in our place. (C1)

An understanding of the importance of MNR to conservation also emerged in Q3 (positive changes), for example:

> Environmentally we are happy as the community because there are certain tree species which are being preserved while they are in there... So in this way the generation will be able to see those species. So that is a plus in terms of environmental issues. Same with animal species. There are those species which you can see like your zebras and other animals, so that is a plus because ...you may have noticed, people in the villages are not concerned about environmental issues. ...it's more about employment. But it is also very important for us as the community to balance the two... (C1)

A senior figure's response to the above-mentioned quote also highlights conservation of the environment as a positive move: "It's ...very good, she's got a very good point, of which people, some of us, don't understand. We only want what's there" (C1).

Community projects

Emerging strongly from the data is the impressive array of community projects run by the reserve, some in partnership with other organisations. Beginning with Q3 (positive changes wrought by reserve), C2 participants in particular expound on several community projects, education/training initiatives, employment opportunities and revenue sharing. C1 could be unaware of these or do not view them as positive changes.

Assistance for local schools and sports programmes are mentioned by C2 as aspects that have changed local people's lives in a positive sense. Community projects emerge again from C2 in Q10 where participants were asked to express their dream for an ideal future. In Q4, community projects are the most mentioned aspect that C2 believe increases positivity in the community, with zero mentions from C1 (C1:0; C2:6). In this same question, for C1, development and infrastructure is what would make them most positive, with no mention of this by C2 (C1:10; C2:0). This indicates differing understandings between C1 and C2 on the crucial aspect of what would most increase positivity.

Examples of the numerous comments by C2 recorded here are helping local schools with building materials and fencing; sports tournaments, training and equipment; contract projects such as Working for Water and Working on Fire; training field rangers; and environmental education initiatives in local schools.

> First of all, from the games that I've mentioned to you, from the soccer games ...the [winners] get [a prize] ...from the nature reserve ...kit ...and trophies and medals, they get exposure. They get knowledge ...knowledge of soccer and everything. So we're upgrading sport in a way, but through conservation. (C2)

> The Environmental Schools Competition ...made a big change. Because now the local people also want to do conservation. But before they were not concerned about conservation, they were just seeing conservation as that we don't want them to come and take things from inside the reserve. Now they see [MNR] as a benefit to them. (C2)

Employment

Based on the frequency of responses in Q8 (benefits), access to natural resources and visiting the beach appear to be most important, followed by employment and then learning/training.

Employment emerges as the most significant positive change identified by C2 in Q3 (along with access to natural resources) (C1:1; C2:6). The quotes refer to jobs created through permanent employment, tenders, programmes or temporary contracts which have improved lives, e.g.: "There was a [local] guy who built his house from the money he was getting from the reserve. He was supervising the general assistants in cutting exotics"; "...the reserve has changed the community's lives because there are job opportunities"; "95% of the people working permanently here at Mkhambathi are coming from these villages. ...with the smaller programmes like Working for Water, about 99% are coming from these villages. So those I would say are direct benefits". Although C2 are very aware of these positive changes, C1 do not appear to be, to the same extent.

However, even if C1 do not see it as a positive change, they do acknowledge employment as a benefit in Q8, with almost the same frequency as C2 (C1:7; C2:8). In Q4, the theme of employment emerges with the second highest frequency as something that would increase positivity, and, again is mentioned in almost equal proportion by C1 and C2 (C1:5; C2:4). This concurs with Shibia (2010) who posits that employment in ecotourism operations increases awareness in local communities of the importance of conservation. In the context of what would make locals more positive towards conservation, respondents spoke about the need for employment in a poor area with little other opportunity:

> We want because we are poor. We want to work in there. They must create work.... (C1)

> ... the number of employees is very, very low. (C1)

> ... it's actually a cry of the whole nation: a lack of employment. So if the nature reserve can employ more people on a permanent basis... I think the community will appreciate the reserve ...more than they do [now]. (C2)

> If there can be more opportunities of employment. Even if it's temporary, because there are no factories here, it's only the reserve. ...If we can create a little employment, the people can be more positive. (C2)

In Q8 (Benefits), participants spoke about using local contractors so that the money goes to the community and the various training initiatives, for example:

> When we need to renovate ...the cottages, we take the local contractors. We don't take people from far... We've got a database here with all the contractors that are around. ...when we hire people from conservationist upwards, we take anybody. But if it's a field ranger post or general post, we hire here... Also when we [have] the interviews, it's us and one community member who comes and watches if the interview was fair. (C2)

In Q10 on their dreams for an ideal future, by interrogating the content of the quotes, job opportunities and entrepreneurship emerged from both C1 and C2. C1 expressed their desire for development which will bring jobs; while C2 talked about entrepreneurship opportunities that they want to see realised, such as local people developing tourism initiatives in their villages; more use of locals in maintenance jobs such as fixing fences; and micro opportunities for unskilled employment e.g. selling firewood and assisting fishermen who are tourists. One participant stated:

> To me, the communities around here are in really remote areas. I feel like their gold is nature. That is, conserving these natural resources that they have. To me, it's their gold that they can live on. ...if people can all realise and be responsible for their areas. To me a nature reserve is a sample of what we would like the whole area ...to look like. ...For me, what we practice here, could be something ...happening within this whole area. ...you can have campsites outside there [and] homesteads that will be used by visitors. We could conserve those important plants that are outside there. So that whoever comes in can have a tour of this area. A person comes and [goes to see] huts and beautiful scenic views, and sleeps over within the villages. You can ...enjoy traditional dance and food. This [can be an extension of] the nature reserve... That's what I always think of... Just having a nature reserve on its own and protecting it doesn't mean [anything] to me, because we want to expand these areas. Because really, that's the only good that people [will get]. ... (C2)

Employment is clearly important to both C1 and C2, but with over 6000 households in the villages forming part of the Mkhambathi Land Trust, only a handful can obtain employment in this small reserve.

Training

Training emerged in Q8 where both C1 and C2 mentioned training for new skills (in hospitality, field ranging training, infrastructure maintenance, tour guiding, etc.) as a benefit. Training was also mentioned as an aspect that has changed lives positively (Q3), with mention of fire awareness and teacher training, for example:

> There was also another programme that was running, Wild Coast Programme, which was also a section of ECPTA, where they went to school. ...two teachers from each school from the local villages ... underwent environmental training... (C2)

Q7 related to whether community members had any responsibility towards the reserve and how they felt about this. The majority of comments related to the fact that local people do help to protect the reserve, particularly those who have benefitted in the past from training. Gadd (2005) and Stem et al. (2003) also found that pro-conservation behaviours are more associated with participation in indirect tourism benefits, such as training and empowerment. In this research, these appear to have fostered relationships, resulting in community members actively protecting the reserve or displaying positivity towards it. The field rangers mentioned in the quotes below are from two villages which fall outside of the Mkhambathi Land Trust, making the cooperation even more noteworthy.

> You know, from the past two years or so, we've seen people that will take responsibility like that. Through the Wild Coast Project that was here for about five years.... The project employed field rangers from ...both adjacent [villages] (Ndengane and Amadiba) and also Khanyayo and other villages. They were employed on a contractual basis, and we ...got that relationship working with them. Up to now, the responsibility that they take, when they see fires ...they call us and say, "There's fire inside the reserve". So they keep us alert and maybe if we see smoke, we can contact them and say "Is there any fire? Is it inside? How big is it?" And then we go there prepared... (C2)

Another quote referred to these community members being the "eyes and ears ...in terms of looking after this reserve, because it is their reserve" (C2). A C2 participant who is also from the community stated:

> Most of them they do care [about the reserve], but there are a few of them whom are saying, "no, it's for them, not for us", but [only a] few. Most of the community members do understand, since the Wild Coast Project has tried to involve them more in the reserve. Because there was funding, they were getting tenders, they were getting trained, the schoolchildren were having the schools' competition and slept over and took a tour inside the reserve. They do understand. Even the young people now, are understanding what's going on in the reserve. (C2)

Although the quotes come from C2, they suggest that training/capacity building builds relationships that invoke protective behavioural actions amongst locals, even after tangible benefits have ended.

Development and infrastructure

In Q3 on positive changes, there is acknowledgement that the government has made infrastructural improvements due to the presence of MNR: "I think it has improved the standard of living of people because the government have to maintain [it] so that it looks attractive" (C1). However, in Q4, development and infrastructure were highlighted as aspects that would most increase positivity towards MNR, hinting at a current lack of development and infrastructure. Ten quotes emerged here, all from C1, and expressed the desire for improved facilities (for locals and tourists), which could result in more tourists, thus bringing further job opportunities. Specific mention is made of shops, more accommodation options, road access and community facilities such as a clinic and cellular network. For example,

> We like to have that nature reserve. The only thing that must be improved is infrastructure. The infrastructure is very poor here. So that we can have more tourists and young people will get more jobs.
>
> I think if the reserve or Eastern Cape Parks can try to get the investor who's going to make more repairs so that there can be more visitors, that would also create the opportunities of jobs to communities surrounding the reserve.

The desire to see Mkhambathi revived also emerged among the responses to Q4, for example: "the standard of Mkhambathi is very low [compared] to the previous years. Like there was a clinic there, a hospital there. Even the [tourist accommodation is] of very poor standards". The frustration due to lack of development and resultant lack of employment was also mentioned under Q5: "People were expecting employment opportunities which didn't come. ...But the ultimate goal, people want to have Mkhambathi developed so that they can get employment opportunities. I think that's a big thing". Respondents also voiced their desire for improving and upgrading the reserve in Q10 in conjunction with the dream for increased future employment.

It would appear that locals would like to see a thriving reserve as well as have facilities that meet their needs such as a clinic, and that these could improve positivity.

Communication and information sharing

In Q1 on the relationship between the reserve and community, the issue of communication arose. Several C1 and C2 respondents mentioned that the Board of Trustees (comprising community members) does not report back sufficiently to the community. This disappointment with the board and desire for information from them was also expressed in Q3 (negative changes), Q4 (aspects that would encourage positivity towards the reserve) and Q5 (others views of the nature reserve). The board is responsible for reporting back to the community regarding the co-management of the reserve, and for determining how revenue share accruing to the community should be spent. This emerged as a strong theme, which can negatively affect conservation related attitudes. Mosimane and Silva (2015), in their study in Namibia, also found that locals actively sought information from the conservancy management committee without success, leading to dissatisfaction. Nault and Stapleton (2011) identify "sustained support from trusted community leaders" as essential for long-term viable community-based ecotourism development in remote areas. If managed correctly, the potential of the co-management agreement between MNR and the Board of Trustees is immense. Lucas, Brooks, Darnton, and Jones (2008) point to devolving power to ground-level agencies as the most effective way of encouraging pro-environmental behaviour! Some examples of this problem at Mkhambathi and its consequences follow:

> I don't think the board ...is transparent. So I think if the board can be transparent with everything to the community, it could be much better. (C1)

> We are not okay about Mkhambathi because we are [not] getting information (which has been hidden by the Mkhambathi Land Trust Board), so we don't know what's going on in our reserve... (C1)

A comment occurring under Q6 suggests that just knowing what goes on within MNR can improve attitudes and even behaviour, without any direct financial benefits. What makes the statement more pertinent is that it relates to the Amadiba community which is not one of the seven villages benefitting from the trust:

> ... we've also got informers at Amadiba that report to us when there's a fire... . Even when the poachers are getting in, they phone and say, tell your guys to go and watch them... Even the chiefs. I remember the time they were warning the people and telling them they mustn't come and make trouble with the reserve because the reserve belongs to them. Because they are part of the forum. They come to the meetings, so they know everything that happens inside the reserve. (C2)

Finally, this theme recurs under Q10 on dreams for the future, where two C1 participants dream of increased communication with the reserve, e.g. "More communication between Mkhambathi and us, so that we can understand each other better".

In summary, transparent reporting structures by the community members on the board, sharing of information and simply knowing what is going on in the reserve appear to assist in encouraging positive attitudes.

Involvement and responsibility

A sense of involvement was another important theme, also identified by Saufi, O'Brien, and Wilkins (2014). From Q7, it would appear that having a responsibility or form of involvement within MNR does positively influence attitudes as well as pro-conservation behaviour. A quote mentioned above addressed local people being protectors even though the tangible benefits had stopped (such as previous training). Matarrita-Cascante et al. (2010) also found that pro-conservation attitudes and behaviour were encouraged by strong local relationships, interaction, open communication and participation. This is supported by Kideghesho et al. (2007) who state that the impact of benefits can be enhanced by regular contact between local people and staff of the protected area. In this study, C1 participants also provided suggestions on how to encourage reserve protection, such as a voluntary clean-up campaign: "I think the community should participate by cleaning without expecting any money, just to clean there". Another C1 participant added: "Yes, and they could, by that, ...know that they feel wanted, they can feel that they are involved". The significance is that a simple suggestion of goodwill towards the reserve *from the community*, could encourage a feeling of involvement *within the community*. It thus appears that simple, easy to implement projects could positively impact attitudes.

Regarding what would make locals more positive (Q4), a C1 participant stated "I should think that maybe ...the community should be more involved in everything that is happening in the nature reserve". A C2 respondent touched on capacity building and skills transfer (as co-management of the reserve occurs), as a positivity builder :

> You see, this partnership that we have with the communities, whilst we manage together there is also ...a skills transfer (in terms of how to manage your areas). [The board] are there as representatives of the communities to make sure that the communities are not robbed [and] get what they are supposed to get. But in the process they will also be getting [an] understanding of why we are doing this, why it is important and how they can also have some input in terms of this. [Because], within the core management [agreement], there is the issue of capacity building in terms of leadership, management. So there will be opportunities coming in there. So that's why I'm saying [that] this partnership, on its own, will have a positive impact.

On the flip side, it emerged from C1 that not feeling involved causes a sense of exclusion which can negatively impact attitudes.

Overall, it would appear that C1 is not averse to having responsibility, and that individuals who have been involved in the past do assist in protecting the reserve.

Environmental education and awareness

The theme of environmental education and awareness was raised by C1 and C2. In Q8, respondents referred to training on environmental education/awareness as a current benefit. However, they appear to want more of this, because environmental education/awareness surfaced as aspects that would increase positivity (Q4); and as a dream for the future (Q10). One C1 respondent stated " ...[we need] community awareness about how important it is to have the nature reserve. So that the community will understand the impact that the nature reserve could have in our community...". C2 referred to initiatives already in place that they feel enhance positivity, such as the Environmental Schools Competition and school outings to the reserve. MNR has embarked on extensive environmental education to help locals understand the value of the reserve, and to encourage a conservation ethos within the villages themselves. Reference is also made to environmental education induction programmes for those entering MNR (for grass cutting and piece jobs).

Environmental education thus appears to invoke positive attitudes towards MNR. Stem et al. (2003) also found that a greater awareness/appreciation for nature could favour long-term conservation, and hopefully remain instilled in the community should tourism decline. Waylen et al. (2010), too, found that interventions providing community outreach and conservation education had greater likelihood of successfully changing attitudes than those that did not.

Tourism potential not realised

Conspicuous in its absence are quotes relating to benefits gained from tourism. This lack suggests that the potential of tourism to bring benefits to locals has yet to be adequately realised. Data from both constituencies confirmed this :

> ... we want to change their mind set in terms of looking at things. ... because [when] we talk about poaching, people will say "no, that's the way that we [do things]", but they don't think about any other alternatives in terms of their livelihoods. ... [for example] using [the] beautiful areas ... in their communities. I'm talking about some waterfalls ... and nice scenery within those communities. And that's why we are saying, we would like to have some projects like small campsites that will be managed by communities in the villages... So that at least they can protect their [environment], because all this harvesting of forests would stop because they could see some benefits coming [and realise] the importance of conservation. (C2)

> ... If they can get more training... so that there can be backpackers ... within the community. That can make [community members] positive too. (C2)

The development of tourism both within MNR and surrounding communities requires attention. Literature attests to tourism contributing to and diversifying livelihoods, increasing the likelihood that locals accept resource use restrictions (Stoll-Kleeman, 2005), reducing dependence on natural resources for survival (Gurung & Seeland, 2011) and encouraging environmental consciousness as locals link a healthy environment with benefits (Matarrita-Cascante et al., 2010). Quotes from C1 attest to tourism at Mkhambathi providing more benefits in the past, and a longing for such success to return.

Overall results

Overall, for Q2, C2 (which constituted five interviewees) were considerably more vocal and positive regarding the relationship between the local community and the reserve, mentioning several actions taken by C2 which positively affect this relationship. In this same vein of positivity, C2 also mentioned far more positive changes (Q3) than C1, which was a much bigger group, while C1 acknowledged considerably more losses than C2 did. It is interesting though that regarding benefits (Q8), C1 were very expressive, mentioning many. Based on the number of NGT notes, it would appear that these benefits are acknowledged and appreciated by C1. Overall, C1 identify more benefits and positive changes than they do losses and negative changes, which is encouraging.

From a methodological perspective, the volume of NGT notes (quotes) produced by C1 (Q8:52; Q9:44) bodes well for future usage of this technique in rural settings. Participants, especially younger females, were highly responsive, whereas they tended to say less during FGI.

Conclusion and recommendations

This research explored the influences on the attitudes of the Khanyayo Community towards Mkhambathi Nature Reserve. The wider context of the population explosion and complexities of conserving biodiversity, yet while enabling the livelihoods of poor rural communities was explained. Literature promotes benefit-sharing approaches as a solution for these communities, but researchers differ regarding *what* benefits result in positive attitudes towards the natural environment, and on the linkages between benefits, attitudes and behaviour. This research revealed that some benefits led to positive attitudes, while a few resulted in tangible pro-conservation actions. Furthermore, some losses resulted in negative attitudes.

In the case of MNR and Khanyayo village, both tangible and intangible benefits appear to influence pro-conservation attitudes. Examples of tangible benefits were employment, facilities and infrastructure, access to natural resources and training; while intangibles included enjoyment of reserve, involvement, communication, capacity building and environmental education. Although the community may focus more on the tangibles (particularly employment), it appears important to encourage them to see that intangible benefits add value and quality to life. Two C2 participants alluded to the fact that intangible involvement or tangible benefits can impact actual behaviour, stating respectively that few poachers come from the villages that are involved; and that increased revenue to local people will result in the game being protected.

It thus appears that in this case study, the presence of benefit-sharing does have positive repercussions for the community, and positively influences attitudes towards MNR. In a few cases, benefits were directly linked to positive behaviour. Yet, the benefits are simply not enough, and probably, in a small reserve with a shoestring budget and large population on its doorstep, they never will be sufficient.

However, there are lessons learnt from this research, particularly from comparing C1 and C2, which yield insight into manageable changes that could be made by protected area management which may impact on pro-conservation attitudes and successful benefit-sharing. These are discussed below.

Resource use is important and appreciated. This is acknowledged by both C1 and C2. Where access is not allowed or restricted, C1 feel this loss sorely, but this did not emerge from C2. It is important for reserve management to take cognisance and acknowledge the losses felt by locals. These are complex issues, but could be partly addressed through environmental education (e.g. on sustainable harvesting of medicinal plants outside MNR), environmental awareness (e.g. on why natural resources within MNR require conservation) and provision of information (e.g. how the fishing restrictions work and how to obtain permits). It is probable that this would be positively received by the community, as they requested more environmental education initiatives.

It is encouraging that the community also enjoy and use their reserve, and this appears to boost positive attitudes. The fact that most participants had entered the reserve is unusual in the African context. They were highly positive regarding their experience, and the authors have not seen evidence of this in similar studies. Based on C2 mentioning this less and ranking it lower, it could be that C2 underestimate the importance of this benefit to locals. Facilitating more visits to the beach would most likely be appreciated, as this is a highly sought after benefit, and most locals currently go only on New Year's Day when transport is provided.

There appears to be a genuine intention by C2 to employ locals, provide training opportunities and stimulate environmental awareness/education. A range of promising initiatives are already in place. However, these are well known to C2 participants, in contrast to C1 barely acknowledging them. Perhaps C1 are largely unaware of these or do not perceive them as benefits? Either way, this could imply the need for communication to inform locals about what reserve management is doing and the benefits that result from the presence of MNR and these initiatives. According to C2, these opportunities have resulted in local people's lives being changed and now being able to build homes, get jobs and learn skills.

On the other hand, development and infrastructure are highly acknowledged by C1 as something they desire and that would contribute to positivity, but this is not recognised by C2 (who see community projects as the aspect that would most increase positivity towards MNR). It could be that the difference lies in the type of development envisaged by the two constituencies. It may be constructive to hold discussions with the community to interrogate this together. It is also significant that, in something as important as what increases positivity towards the reserve, that the views of C1 and C2 are so polarised. It seems as if C1 need to be made more aware of community projects; as well as C1 and C2 engaging more on the vision for development/infrastructure.

Involvement and interaction is essential, and need not be overly complex. Sometimes a simple project (e.g. a clean-up campaign) can encourage a sense of involvement for locals. Continuous open lines of communication between MNR and the community regarding what is going on can go a long way to improving attitudes towards conservation. C2 were more aware of positive changes than C1. This appears to be a significant gap since knowledge of these positive aspects could influence attitudes towards MNR.

Information provision can extend to clarifying the rules regarding resource use/access, e.g. some C1 participants feel they are not allowed to fish. Reserve management could discuss this at community meetings – fishing zones and how to acquire a permit. Others thought they could no longer access sites of spiritual importance. Providing this information may reduce negativity. The poor communication and lack of transparency of the community members on the Board of Trustees manifests as a barrier to positive attitudes. Reserve management are well aware of this problem.

Environmental education also emerged as a benefit and as something that would increase positivity. Considering the favourable results this has had in other research confirms environmental education as a promising avenue to continue to pursue at Mkhambathi.

The Wild Coast Project surfaced regularly in quotations and proves the value of a well-run programme. It was mentioned as an initiative that built capacity, improved relationships, and positively influences pro-conservation behaviour to this day, even though the project has ceased. Direct positive behaviours from this have been community members reporting escaped animals, fires and poachers. This and other evidence emerging under Q7 suggests that locals have a sense of responsibility towards MNR, and want to be involved, but some feel excluded which impacts on attitudes. In addition, involvement appears to reduce poaching, with less poaching from communities that form part of the Mkhambathi Land Trust.

Finally, it is encouraging that the community see more benefits and positive changes than losses and negative changes; and that they acknowledge conservation as important. MNR has phenomenal natural attractions, yet few tourism facilities. The reserve is surrounded by very poor communities, with little options of employment apart from the reserve. The outstanding natural beauty of MNR holds significant potential for tourism. With the current private sector development in a portion of the reserve (for high-end tourism) and if the long-awaited government attention to the area materialises, or funding is procured to repair the unused accommodation units, tourism numbers could increase. In high season, the three ECPTA units are fully booked (and long in advance). Tourist demand exists, but supply needs to match up. Increased tourism is essential to provide the impetus for a conservation ethos to expand into the community which has its own natural resources which require conservation and can be shared with tourists, creating micro-employment opportunities. More tourism would also provide additional employment within MNR, both in the formal and informal sector.

The well-known adage, "Conservation without people is just conversation" is a reminder that the people are vital, and for people to support parks requires good mutually beneficial relationships. Although several findings corroborate with those from other studies (Gadd, 2005; Kideghesho et al., 2007; Matarrita-Cascante et al., 2010; Mosimane & Silva, 2015; Saufi et al., 2014; Shibia, 2010; Stem et al., 2003; and Waylen et al., 2010) the comparison between C1 and C2 provides interesting insights into where the constituencies concur and differ. It is widely accepted that benefit-sharing is important. However, this research has demonstrated the value of establishing where the perceptions of benefits differ according to people (C1) and parks (C2), so that through practical actions, these can be aligned to improve the relationship. In addition, understanding which benefits matter to the community, and how these can be built on to improve lives is essential. This alignment is a step towards protected areas surviving and local people thriving.

Notes

1. Statistics South Africa (2011) define village as a settlement in a tribal area with boundaries that delimit populated areas as well as agricultural areas such as grazing land, crop land and forests. Villages usually fall under the jurisdiction of tribal authorities headed by chiefs.
2. The reserve name used to be spelt as "Mkambati", but was changed to "Mkhambathi" after 1994. Both versions are, however, still found in literature regarding the reserve.
3. More recent statistics based on the 2011 national census are unreliable.
4. This research forms part of a PhD where various protected areas and the communities closest to them were selected as case studies.

5. The mapping results are beyond the scope of this article and are therefore excluded.
6. Traditional African medicine, often derived from trees and other plants, and believed to cure various ailments.

Acknowledgments

The authors gratefully acknowledge the Khanyayo Chief, Khanyayo village participants and the reserve staff who participated in this research. The reserve manager, Mr Vuyani Mapiya is thanked for his invaluable support, and Prof. Ciné van Zyl for her input as co-supervisor of the PhD study from which this research originates. Thank you to the University of South Africa for funding provided through its Academic Qualification Improvement Programme.

Disclosure statement

No potential conflict of interest was reported by the authors.

ORCID

Dorothy Queiros http://orcid.org/0000-0002-6988-5818
Kevin Mearns http://orcid.org/0000-0001-5874-3542

References

Abrams, R. W., Anwana, E. D., Ormsby, A., Dovie, D. B. K., Ajagbe, A., & Abrams, A. (2009). Integrating top-down with bottom-up conservation policy in Africa. *Conservation Biology, 23*(4), 799–804. doi:10.1111/j.1523-1739.2009.01285.x

ATLAS.ti. Version 7.0 [Computer Software] (2012-2016). Berlin: Scientific Software Development. Retrieved fromhttp://atlasti.com/.

Baker, A. (2015). Let's prepare for Africa's population surge now or face the consequences. *TIME, June 22, 44*.

Barbour, R. S. (2014). Quality of data analysis. In U. Flick (Ed.), *The SAGE handbook of qualitative data analysis* (pp. 496–509). London: SAGE Publications Ltd.

Bennett, N. J., & Dearden, P. (2014). Why local people do not support conservation: Community perceptions of marine protected area livelihood impacts, governance and management in Thailand. *Marine Policy, 44*, 107–116. doi:10.1016/j.marpol.2013.08.017

Berkes, F. (2004). Rethinking community-based conservation. *Conservation Biology, 18*(3), 621–630. doi:10.1111/j.1523-1739.2004.00077.x

Blore, M. L., Cundill, G., & Mkhulisi, M. (2013). Towards measuring the transaction costs of co-management in Mkambati Nature Reserve, Eastern Cape, South Africa. *Journal of Environmental Management, 129*, 444–455. doi:10.1016/j.jenvman.2013.08.002

Burgoyne, C., & Mearns, K. (2017). Managing stakeholder relations, natural resources and tourism: A case study from Ololosokwan, Tanzania. *Tourism and Hospitality Research, 17*(1), 68–78. doi:10.1177/1467358416639088

Campbell, L. M., Haalboom, B. J., & Trow, J. (2007). Sustainability of community-based conservation: Sea turtle egg harvesting in Ostional (Costa Rica) ten years later. *Environmental Conservation, 34*(2), 122–131. doi:10.1023/A:1025071822388

Chapple, M., & Murphy, R. (1996). The nominal group technique; extending the evaluation of students' teaching and learning experiences. *Assessment and Evaluation in Higher Education, 21*(2), 147–159. doi:10.1080/0260293960210204

Cohen, L., Manion, L., & Morrison, K. (2011). *Research methods in education* (seventh). New York, NY: Routledge.

Coles, D., Duval, D. T., & Shaw, G. (2013). *Student's Guide to writing dissertation and theses in tourism studies and related disciplines*. Abingdon: Routledge.

Coria, J., & Calfucura, E. (2012). Ecotourism and the development of Indigenous communities: The good, the bad, and the ugly. *Ecological Economics, 73*, 47–55. doi:10.1023/A:1025071822388

Creswell, J. W. (2014). *Research Design: Qualitative, quantitative and mixed methods approaches* (Fourth). California : SAGE Publications Ltd.

Darbyshire, P., Macdougall, C., & Schiller, W. (2005). Multiple methods in qualitative research with children: more insight or just more?. Qualitative Research, *5*(4), 417–436, 10.1177/1468794106058867.

Davies, T. E., Fazey, I. R. A., Cresswell, W., & Pettorelli, N. (2014). Missing the trees for the wood: Why we are failing to see success in pro-poor conservation. *Animal Conservation, 17*(4), 303–312. doi:10.1023/A:1025071822388

de Los Angeles Somarriba-Chang, M., & Gunnarsdotter, Y. (2012). Local community participation in ecotourism and conservation issues in two nature reserves in Nicaragua. *Journal of Sustainable Tourism, 20*(8), 1025–1043. doi:10.1080/09669582.2012.681786

Fihlani, M. (2015). *Integrated development plan review, 2015/2016*. Ingquza Hill Local Municipality: Flagstaff.

Gadd, M. E. (2005). Conservation outside of parks: Attitudes of local people in Laikipia, Kenya. *Environmental Conservation, 32*(1), 50–63. doi:10.1017/S0376892905001918

Gibbs, G. R. (2014). Using software in qualitative analysis. In U. Flick (Ed.), *The SAGE handbook of qualitative data analysis* (pp. 277–294). London: SAGE Publications Ltd.

Grbich, C. (2007). *Qualitative data analysis: An introduction*. London: SAGE Publications Ltd.

Gubrium, J. F., & Holstein, J. (2014). Analytic inspiration in ethnographic fieldwork. In U. Flick (Ed.), *The SAGE handbook of qualitative data analysis* (pp. 35–48). London: SAGE Publications Ltd.

Gurung, D. B., & Seeland, K. (2011). Ecotourism benefits and livelihood improvement for sustainable development in the nature conservation areas of Bhutan. *Sustainable Development, 19*(5), 348–358. doi:10.1002/sd.443

Imran, S., Alam, K., & Beaumont, N. (2014). Environmental orientations and environmental behaviour: Perceptions of protected area tourism stakeholders. *Tourism Management, 40*, 290–299. doi:10.1016/j.tourman.2013.07.003

Karanth, K. K., & Nepal, S. K. (2012). Local residents perception of benefits and losses from protected areas in India and Nepal. *Environmental Management, 49*(2), 372–386. doi:10.1007/s00267-011-9778-1

Kepe, T. (2004). Land restitution and biodiversity conservation in South Africa: The case of Mkambati, Eastern Cape province. *Canadian Journal of African Studies, 38*(3), 688–704. doi:10.1080/00083968.2004.10751304

Kepe, T. (2007). Medicinal plants and rural livelihoods in Pondoland, South Africa: Towards an understanding of resource value. *International Journal of Biodiversity Science and Management, 3*(3), 170–183. doi:10.1080/17451590709618171

Kepe, T. (2008). Land claims and comanagement of protected areas in South Africa: Exploring the challenges. *Environmental Management, 41*(3), 311–321. doi:10.1007/s00267-007-9034-x

Kepe, T. (2012). Land and justice in South Africa : Exploring the ambiguous role of the state in the land claims process. *African and Asian Studies, 11*(4), 391–409.

Kideghesho, J. R., Røskaft, E., & Kaltenborn, B. P. (2007). Factors influencing conservation attitudes of local people in Western Serengeti, Tanzania. *Biodiversity and Conservation, 16*(7), 2213–2230. doi:10.1007/s10531-006-9132-8

Kiss, A. (2004). Is community-based ecotourism a good use of biodiversity conservation funds ? *Trends in Ecology and Evolution, 19*(5), 232–237. doi:10.1016/j.tree.2004.03.010

Langholz, J. (1999). Exploring the effects of alternative income opportunities on rainforest use: Insights from Guatemala's Maya Biosphere Reserve. *Society and Natural Resources, 12*, 139–149. doi:10.1080/089419299279803

Lee, T. H. (2013). Influence analysis of community resident support for sustainable tourism development. *Tourism Management, 34*, 37–46. doi:10.1016/j.tourman.2012.03.007

Liu, J., Qu, H., Huang, D., Chen, G., Yue, X., Zhao, X., & Liang, Z. (2014). The role of social capital in encouraging residents' pro-environmental behaviours in community-based ecotourism. *Tourism Management, 41*, 190–201. doi:10.1016/j.tourman.2013.08.016

Lucas, K., Brooks, M., Darnton, A., & Jones, J. E. (2008). Promoting pro-environmental behaviour: Existing evidence and policy implications. *Environmental Science and Policy, 11*, 456–466. doi:10.1016/j.envsci.2008.03.001

Matarrita-Cascante, D., Brennan, M. A., & Luloff, A. E. (2010). Community agency and sustainable tourism development: The case of La Fortuna, Costa Rica. *Journal of Sustainable Tourism, 18*(6), 735–756. doi:10.1080/09669581003653526

Mbaiwa, J. (2005). Wildlife resource utilization at Moremi Game Reserve and Khwai community area in the Okavango Delta, Botswana. *Journal of Environmental Management, 77*(2), 144–156. doi:10.1016/j.envman.2005.03.007

McShane, T. O., Hirsch, P. D., Trung, T. C., Songorwa, A. N., Kinzig, A., Monteferri, B., & Mutekanga, D. (2011). Hard choices: Making trade-offs between biodiversity conservation and human well-being. *Biological Conservation, 144*(3), 966–972. doi:10.1016/j.biocon.2010.04.038

Mehta, J. N., & Heinen, J. T. (2001). Does community-based conservation shape favorable attitudes among locals? An empirical study from Nepal. *Environmental Management, 28*(2), 165–177. doi:10.1007/s002670010215

Miles, B. M., Huberman, A. M., & Saldaña, J. (2014). *Qualitative data analysis: A methods sourcebook* (3rd ed.). California : SAGE Publications, Inc.

Mosimane, A. W., & Silva, J. A. (2015). Local governance institutions, CBNRM and benefit-sharing systems in Namibian conservancies. *Journal of Sustainable Development, 8*(2), 99–112. doi:10.5539/jsd.v8n2p99

Moswete, N., Thapa, B., & Child, B. (2012). Attitudes and opinions of local and national public sector stakeholders towards Kgalagadi Transfrontier Park, Botswana. *International Journal of Sustainable Development and World Ecology, 19*(1), 67–80. doi:10.1080/13504509.2011.592551

Nault, S., & Stapleton, P. (2011). The community participation process in ecotourism development: A case study of the community of Sogoog, Bayan-Ulgii, Mongolia. *Journal of Sustainable Tourism, 19*(6), 695–712. doi:10.1080/09669582.2010.536240

Nyaupane, G. P., & Poudel, S. (2011). Linkages among biodiversity, livelihood, and tourism. *Annals of Tourism Research, 38* (4), 1344–1366. doi:10.1016/j.annals.2011.03.006

Patton, M. (2005). *Encyclopedia of statistics in behavioral science*. New Jersey: John Wiley & Sons, Inc. doi:10.1002/0470013192.bsa514

Palmberger, M., & Gingrich, A. (2014). Qualitative comparative practices: Dimensions, cases and strategies. In U. Flick (Ed.), *The SAGE handbook of qualitative data analysis* (pp. 94–108). London: SAGE Publications Ltd.

Queiros, D. R. (2000). *Implementing the fundamentals of ecotourism: The case study of Mkambati Nature Reserve, Wild Coast, South Africa* (Master's thesis). University of Pretoria, Pretoria.

Reimer, J. K. K., & Walter, P. (2013). How do you know it when you see it? Community-based ecotourism in the Cardamom Mountains of southwestern Cambodia. *Tourism Management, 34*, 122–132. doi:10.1016/j.tourman.2012.04.002

Roulston, K. (2014). Analysing interviews. In U. Flick (Ed.), *The SAGE handbook of qualitative data analysis* (pp. 297–312). London : Sage Publications Ltd.

Saufi, A., O'Brien, D., & Wilkins, H. (2014). Inhibitors to host community participation in sustainable tourism development in developing countries. *Journal of Sustainable Tourism, 22*(5), 801–820. doi:10.1080/09669582.2013.861468

Saunders, M., Lewis, P., & Thornhill, A. (2009). *Research Methods for Business Students* (5th ed.). Harlow: Pearson Education Limited.

Shibia, M. G. (2010). Determinants of attitudes and perceptions on resource use and management of Marsabit National Reserve, Kenya. *Journal of Human Ecology, 30*(1), 55–62.

Simpson, M. C. (2008). Community benefit tourism initiatives – A conceptual oxymoron ? *Tourism Management, 29*, 1–18. doi:10.1016/j.tourman.2007.06.005

Snyman, S. L. (2012). The role of tourism employment in poverty reduction and community perceptions of conservation and tourism in southern Africa. *Journal of Sustainable Tourism, 20*(3), 395–416. doi:10.1080/09669582.2012.657202

Snyman, S. (2014). Assessment of the main factors impacting community members' attitudes towards tourism and protected areas in six southern African countries. *Koedoe, 56*(2), 1–12. doi:10:4102/koedoe.v5612.1139

South Africa. Statistics South Africa. (2011). *Census 2011: Metadata* (Report No. 03-01-47). Pretoria: Statistics South Africa. Retrieved fromhttp://www.statssa.gov.za/census/census_2011/census_products/Census_2011_Metadata.pdf.

Steg, L. & Vlek, C. (2009). Encouraging pro-environmental behaviour: An integrative review and research agenda. *Journal of Environmental Psychology, 29*(3), 309–317. doi:10.1016/j.jenvp.2008.10.004

Stem, C. J., Lassoie, J. P., Lee, D. R., David, D., Schelhas, J. W., Deshler, D. D., & Schelhas, J. W. (2003). Community participation in ecotourism benefits: The link to conservation practices and perspectives. *Society & Natural Resources, 16*(5), 387–413. doi:10.1080/08941920390190041

St John, F. A. V., Edwards-Jones, G., & Jones, J. P. G. (2010). Conservation and human behaviour: Lessons from social psychology. *Wildlife Research, 37*(8), 658–667. doi:10.1071/WR10032

Stoll-Kleemann, S. (2005). Voices for biodiversity management in the 21st century. *Environment, 47*(10), 24–36. doi:10.3200/ENVT.47.10.24-36

Stronza, A., & Gordillo, J. (2008). Community views of ecotourism. *Annals of Tourism Research, 35*(2), 448–468. doi:10.1016/j.annals.2008.01.002

Torquebiau, E., & Taylor, R. D. (2009). Natural Resource Management by rural citizens in developing countries: Innovations still required. *Biodiversity and Conservation, 18*(10), 2537–2550. doi:10.1007/s10531-009-9706-3

Tran, L., & Walter, P. (2014). Ecotourism, gender and development in northern Vietnam. *Annals of Tourism Research, 44*(1), 116–130. doi:10.1016/j.annals.2013.09.005

Van Tejilingen, E., Pitchforth, E., Bishop, C., & Russell, E. (2006). Delphi method and nominal group technique in family planning and reproductive health research. *Journal of Family Planning and Reproductive Health care, 32*(4), 249–252. doi:10.1783/147118906778586598

Waylen, K. A., Fischer, A., Mcgowan, P. J. K., Thirgood, S. J., & Milner-Gulland, E. J. (2010). Effect of local cultural context on the success of community-based conservation interventions. *Conservation Biology, 24*(4), 1119–1129. doi:10.1111/j.1523-1739.2010.01446.x

Zeka, S. (2011). Intricacies of land restitution in Africa. *Aisa Policy Brief, 34*, 1–5.

Strengthening governance processes to improve benefit-sharing from tourism in protected areas by using stakeholder analysis

Jasper Heslinga, Peter Groote and Frank Vanclay

ABSTRACT

In tourism, the concept of "benefit-sharing" refers to the idea that the benefits arising from tourism should be distributed across a wide range of stakeholders. We argue that the development of synergetic interactions between stakeholders involved in governance processes is a prerequisite for effective benefit-sharing from tourism in protected areas. Our stakeholder analysis of the actors with an interest in the island of Terschelling in the northern Netherlands revealed how relationships between stakeholders enable and/or constrain the sharing of benefits from tourism. Our analysis helped to understand the governance arrangements pertaining to the management of tourism in protected areas. We ascertained that the national forest management agency (*Staatsbosbeheer*), a large landowner on the island, is highly influential, but nevertheless often found it difficult to gain local support for its activities. The local government was also an important stakeholder, but was considered to sometimes constrain the development of tourism and thus limit the potential for benefit-sharing. Effective communication, good collaboration with stakeholders, and an attitude of openness were identified as being important preconditions for developing synergistic interactions between stakeholders.

Introduction

We consider that "benefit-sharing" in the context of tourism in protected areas refers to the idea that the benefits arising from tourism should be shared amongst a wide range of stakeholders, and especially with local communities (Foxlee, 2007). Despite widespread use of the term, benefit-sharing (Söderholm & Svahn, 2015; Vanclay, 2017; Wang, 2012), there is no well-established definition (Swemmer, Grant, Annecke, & Freitag-Ronaldson, 2015). However, in the context of protected areas, benefit-sharing can be defined as being "the process of making informed and fair trade-offs between social, economic and ecological costs and benefits within and between stakeholder groups, and between stakeholders and the natural environment, in a way that is satisfactory to most parties" (Swemmer et al., 2015, p.7).

In this paper, we move beyond "trade-offs" to argue that "synergies" can be an important precursor to achieve effective benefit-sharing from tourism in protected areas. Developing synergies can be described as facilitating the interactions between actors to achieve greater combined outcomes across the social and ecological domains (Persha, Agrawal, & Chhatre, 2011). For example, this could mean that tourism development and nature protection should not be considered as being in conflict, but rather as goals that can be balanced to create win-win situations and be mutually supportive.

Acknowledging synergetic tourism–landscape interactions is important because tourism generates income and job opportunities that rely on the landscape, although simultaneously tourism impacts on the surrounding landscape (Buckley, 2012; Saarinen, 2006) and the community (King, Pizam, & Milman, 1993; McCombes, Vanclay, & Evers, 2015; Snyman, 2015). Tourism is also an opportunity for nature protection, because tourism plays a role in creating awareness, public support and in generating funding for nature protection (Libosada, 2009; McCool & Spenceley, 2014). To achieve both nature protection and socio-economic development, we consider it is important that the synergetic interactions between tourism and protected areas be recognized and stimulated. In other words, instead of fair trade-offs in a zero-sum game, combining both objectives can create synergetic effects, win-win-win outcomes, as well as enhancing benefit-sharing possibilities.

A problem, however, is that the potential synergies between tourism development and landscape protection (T&L) are often overlooked and underutilized (Hartman, 2015; Heslinga, Groote, & Vanclay, 2017). One way to identify synergies lies in the governance arrangements in managing T&L in coastal areas (Lockwood, 2010). This is because these governance arrangements affect the processes by which synergies are activated or inhibited. In this paper, we aim to improve the understanding of the governance processes that could help facilitate benefit-sharing arising from tourism in protected areas.

Surprisingly, governance as a conceptual frame has only had relatively limited use in the tourism literature (Bramwell & Lane, 2011). To make up for this, we explore governance from the perspective of social–ecological systems (SES) theory. SES theory is useful because it does not consider the social and the ecological as distinct separate entities, but instead as an integrated whole. Using a holistic approach helps in understanding the interactions between tourism and landscape protection in coastal areas (Heslinga et al., 2017). In this paper, we draw on an SES perspective to provide principles that contribute to a deeper understanding of how tourism–landscape interactions in coastal areas can be better managed.

To understand governance arrangements better, examining the relationships between the stakeholders can be helpful. We use stakeholder analysis because it can reveal the interests and influences of the different stakeholders, and determine whether their interactions are conflicting, complementary or cooperative (Reed et al., 2009). It is generally accepted that the use of real-life examples assists in demonstrating the usefulness of an approach (Wesley & Pforr, 2010). Therefore, we utilize the case of Terschelling, an island located in the Wadden Sea region, a UNESCO World Heritage site in the north of the Netherlands. The Wadden is renowned for its ecological qualities and highly valued landscapes. Due to its attractiveness, tourism is a significant activity, especially on the five main islands of the Dutch Wadden. On Terschelling, there are many stakeholders groups who are involved in decision-making processes related to tourism, each with their own varying interests. Our stakeholder analysis identified who these stakeholders were and helped understand the interactions between them.

Governance to facilitate benefit-sharing from tourism in protected areas

Despite the potential for synergetic tourism–landscape interactions in coastal areas, managing these interactions can be inherently complicated because many actors are involved in decision-making processes and these stakeholders usually have different and sometimes contradictory values, attitudes and interests. What an environmental stakeholder, for example, considers to be important in an area can be very different to what tourism promoters, developers, recreational users, or local residents consider to be important (Jamal, 2004). Given the complexity of the stakeholder interactions and the difficulties of managing their diverse interests, increasing attention has been given to managing tourism development processes (Luthe & Wyss, 2014; Wesley & Pforr, 2010). The concept of governance, for example, is a promising approach for managing synergies between tourism and landscape in coastal areas. This is because "governance" is a broader concept than "the government", in that it also includes non-state actors, including business, community and civil society, notably the voluntary sector (Parra, 2010). Governance can be defined as "the complex system of regulation involving the interactions of a wide variety of actors, institutions, the environment and all types of socio-institutional arrangements at different territorial levels" (Parra, 2010, p. 491).

Balancing the objectives of nature protection and socio-economic development, and thereby achieving long-term sustainability goals, requires organizational structures that are more decentralized than central governments tend to be, as well as effective linkages between the many stakeholders (Crona & Bodin, 2006; Imperiale & Vanclay, 2016; Plummer & Fennel, 2009; Reed et al., 2009; Strickland-Munro, Allison, & Moore, 2010). Central governments can be useful in assisting in the formation of groups and in providing support for collective action, but they sometimes interfere when there is a well-functioning civil society (Mehmood & Parra, 2013). Governance arrangements that accommodate inclusion and participation are desirable so that effective rules, institutions and incentives can be developed to influence the management of tourism–landscape interactions in a complex and uncertain world (Armitage et al., 2009).

The interest in governance as a concept has increased significantly in the social sciences over the last few decades (Bramwell & Lane, 2011; Kooiman, 2003; Wray, 2015). Nevertheless, despite the potential of governance to provide insights regarding the management of tourism–landscape interactions, its usage in the tourism literature is limited. Bramwell and Lane (2011) claim that the term "governance" has been used less frequently than related terms – e.g. tourism politics, policy, policy-making, planning, or destination management. When Bramwell and Lane (2011) wrote their article, only a few scholars working on governance in relation to tourism were influential (Eagles, 2009; Hall, 2011; Higgins-Desbiolles, 2011; Wesley & Pforr, 2010). Since then, usage of governance in relation to tourism has increased considerably (Halkier, 2014; Sharpley & Ussi, 2014; Wray, 2015).

The concept of governance has been discussed at some length in other bodies of literature, notably in SES theory (Brondizio, Ostrom, & Young, 2009). Therefore, to enrich the tourism literature, we connect with SES theory. We believe that an SES perspective can be useful to understand tourism–landscape interactions in coastal areas (Heslinga et al., 2017). This is because such a perspective sees tourism and landscape as part of an integrated social and ecological system. Additionally, an SES perspective helps to understand T&L as part of a complex SES that is continually adapting to changing circumstances (Strickland-Munro et al., 2010).

For the management of tourism destinations to address the twin goals of nature conservation and socio-economic regional development, it is important to understand how SESs are governed and to consider the roles institutions can, do and could play (Bramwell & Lane, 2011). SES thinking provides principles for the way tourism–landscape interactions should be governed. In our previous work (Heslinga et al., 2017), we identified three principles – inclusiveness, more flexible social arrangements, and multi-scalarity – which we use as an organizing structure for this paper. These principles are explained below.

Inclusiveness is a principle around the ideas that all actors have a right to be involved in the decision-making process, that they should be given every opportunity to be involved, and that no actors are excluded (Lockwood, 2010). However, the relevant actors and stakeholders involved in the governance of T&L are often diverse and have varied interests and priorities (Bramwell & Lane, 2011; Jamal & Stronza, 2009). Additionally, the interactions between tourism and landscape span numerous policy domains. These characteristics make effective decision-making complex. Including all the different interests fairly, and avoiding the marginalization of any group, can help to prevent conflict among stakeholders (Prenzel & Vanclay, 2014). Conflict may arise in the governance of tourism–landscape interactions because each group is likely to pursue their preferred policy outcomes (Bramwell & Lane, 2011). Avoiding conflict is crucial, because it can impact on economic, ecological and socio-cultural well-being (Jamal & Stronza, 2009).

Another principle of good governance is flexible social arrangements. Governance includes more actors than just the central government. Although centralized government bodies can be helpful in building support for collective action (Olsson, Folke, & Hahn, 2004; Prell, Hubacek, & Reed, 2009), a central government is often limited in its ability to respond to rapid social-ecological change or to cope with uncertainty (Armitage et al., 2009). We believe that governance arrangements in which flexibility is taken into account can be helpful in developing more effective ways of managing tourism–landscape interactions in a dynamic world. An adaptive approach that is flexible enough to deal with future social and ecological changes can assist in enabling progressive learning at individual, community, institutional, and policy levels (Plummer & Armitage, 2007).

Multi-scalarity implies that the governance processes of T&L do not only take place at one single level (or scale), but are also influenced from multiple scales (Adger et al., 2003; Berkes, 2007; Brondizio et al., 2009; Folke, Hahn, Olsson, & Norberg, 2005; Lew, 2014; Liu et al., 2007). The difficulties of managing T&L include discrepancies between socio-economic activities occurring at the local scale and nature protection initiatives imposed from higher scales. Paloniemi and Tikka (2008) noted, for example, that nature protection has been prescribed and stipulated in international and national laws and procedures regulated by the public sector. However, they also observed that the issues and relationships surrounding nature protection play out in various ways at the local level. The everyday lives of local people are affected by their social positions, cultural activities and cultural heritage (Vanclay, 2012). For the management of tourism–landscape interactions, this could mean a mismatch between the regulations for nature protection and those for the socio-economic activities of tourism entrepreneurs (Paloniemi & Tikka, 2008). Acknowledging these multi-scalar tourism–landscape interactions can help to understand the difficulties in managing them better.

Despite the interesting insights SES theory provides for understanding the governance of tourism–landscape interactions in coastal areas, there is a need for more diverse and more detailed case studies into coastal tourism development planning and management (Wesley & Pforr, 2010). These real-life cases can help demonstrate the usefulness of SES. We suggest that using stakeholder analysis helps understand governance arrangements better, as it illuminates the action arena within these arrangements by revealing each actor's positions, interests, and how they interact with others (Ostrom, 2011).

Data and methods for our case study

We used stakeholder analysis to identify the stakeholders, consider their interests, and to analyze whether their inter-relationships were conflictual, complementary, or based on cooperation. Reed et al. (2009) described stakeholder analysis as a process that identifies the individuals, groups and organizations (including future generations and non-human and non-living entities) who are affected by or can affect a decision, action or part thereof. We used stakeholder analysis because it identifies who the stakeholders are, what their interests are, who has the power to influence what happens, and how the stakeholders interact.

Stakeholder analysis has increasingly been used in many different fields and for an increasing variety of purposes (Reed et al., 2009). From its original application in the field of strategic management (Freeman, 1984), it is now widely used in the fields of policy studies, development studies and natural resource management (Dare, Schirmer, & Vanclay, 2014; Prell et al., 2009; Reed et al., 2009). Stakeholder analysis has often been used in tourism research, especially in relation to sustainable tourism development (Currie, Seaton, & Wesley, 2009; Waligo, Clarke, & Hawkins, 2013). To understand the shared objectives between tourism development and landscape protection, we emphasize the importance of involving all stakeholders. We believe that, in a tourism context, stakeholder analyses that focus solely on the tourism industry are inadequate. A tourism destination is more than just economic activity (Higgins-Desbiolles, 2006); the social (King et al., 1993; McCombes et al., 2015) and environmental aspects (Buckley, 2012; Saarinen, 2006) must also be taken into account. Therefore, we are interested in the multitude of different stakeholders that are involved in the management of T&L in coastal areas, ranging from tourism entrepreneurs, policy-makers, environmental groups, interests groups, and civil society.

There are numerous methods available for analyzing stakeholders and understanding their inter-relationships. We utilized the three steps nominated by Reed et al. (2009): (1) identify the stakeholders; (2) categorize them; and (3) investigate the relationships between them. To identify the stakeholders involved in the decision-making processes on Terschelling, we used a snowballing process as part of the interviews conducted by the lead researcher. To establish the appropriate (number of) stakeholders for any specific analysis is difficult (Eagles, McCool, & Haynes, 2002). To address these difficulties, interviews were used to help scope the stakeholders to be included. Starting from people in our networks, the initial interviews were conducted in early 2015 with well-connected local identities and people knowledgeable about Terschelling or the Dutch Wadden area generally. In the initial eight interviews (which ranged from 60 to 120 min each), a total of some 30 or so stakeholders were identified.

The categorization of these stakeholders was done in August 2016 by a panel comprising tourism researchers from the European Tourism Futures Institute at Stenden University in Leeuwarden, the Netherlands, all of whom were highly knowledgeable about Terschelling. Using a card sorting technique, the panel was asked to discuss and position the identified stakeholders on an interest–influence matrix (Reed et al., 2009, discussed below). This matrix comprises four categories: Key Players, Context Setters, Subjects, and the Crowd. These categories are based on the combination of the amount of interest in and influence the stakeholders have in terms of T&L issues. "Key Players" are those stakeholders who have both a high interest in and high influence over tourism. "Context Setters" are highly influential, but only have little interest. "Subjects" have high interest, but only low influence. They need to form alliances with other stakeholders in order to become more influential. The "Crowd" has little interest and little influence. Each category is depicted by a quadrant in the interest–influence matrix. From a theoretical point of view, the strategies used by a stakeholder should vary according to the category in which they are located.

In order to categorize the stakeholders on Terschelling, the panel was given a set of cards, in random order, on which each identified stakeholder was named. The researchers collectively discussed which quadrant each stakeholder should be best placed and why. The categorization process was audio-recorded and transcribed. This resulted in an interest–influence matrix, which provided an overview of the stakeholders involved in decision-making on Terschelling.

The interrelations between the stakeholders were assessed by a thorough re-analysis of the original eight interviews, augmented by an additional six interviews. For the extra interviews, which were done in September 2016, the interest–influence matrix was used to focus discussion specifically on the interactions between stakeholders. Overall, the 14 interviews (12 men, 2 women) included tourism entrepreneurs, policy-makers, and representatives of environmental interest groups and civil society. Given the nature and topics of the discussion, after 14 interviews, it was considered that saturation had been reached.

The interviews were conducted in a manner consistent with ethical social research principles (Vanclay, Baines, & Taylor, 2013). Prior to the interview, the respondent was provided with a research information sheet and was asked to complete an informed consent form that covered issues of anonymity, the use of the research, and their rights during and after the interview. With the permission of each respondent, the interviews were audio-recorded and transcribed verbatim. The transcripts were coded and analyzed using the qualitative data software, Atlas.ti (version 7.5.10). A priori coding was undertaken with the codes having been derived from our theoretical positioning, especially the principles of inclusiveness, more flexible social arrangements, and multi-scalarity.

The interviews were conducted in Dutch. The transcripts and analysis were done in Dutch. Extracts for this paper were selected and then translated into English by the authors. To preserve the intention and meaning implicit in the Dutch statements, rather than a verbatim, literal translation, some of the excerpts have been modified to ensure that a reader in English comprehends the intended meaning of the statement. We believe we have faithfully represented the essence of the interview in the way the extracts have been translated.

There were some limitations to our methods. For example, we were not able to interview all the stakeholders identified and categorized in the matrix. In addition, the key informants we interviewed often represented an organization – and interviewing a different informant within that organization might have given a different perspective. Nevertheless, despite these limitations, our methods were helpful in providing an illustration of the interactions between stakeholders and different groups of stakeholders.

Case-study description

We examine, at different levels, the governance structure and processes that are in place on Terschelling, an island in the Dutch Wadden region (see Figure 1). We specifically consider the management of tourism–landscape interactions on the island. The Wadden is a natural coastal area that has considerable biodiversity and highly valued landscapes (Kabat et al., 2012). These natural qualities led to the area being designated as a UNESCO World Heritage site in 2009 (Kabat et al., 2012), although amongst some controversy (van der Aa, Groote, & Huigen, 2004). Because of its natural qualities, the Wadden is one of the most popular tourism destinations in the Netherlands (Revier, 2013; Sijtsma, Daams, Farjon, & Buijs, 2012). Tourism to the Wadden islands originally started about 100 years ago. After the Second World War, tourism steadily increased, becoming an important economic activity on the islands (Sijtsma, Broersma, Daams, Hoekstra, & Werner, 2015).

The Wadden can be differentiated into three distinct areas, each with different interactions between tourism and landscape. One area is the Wadden Sea, a tidal mudflat and saltmarsh area that is of considerable ecological importance. Here, a limited amount of tourism activities take place such as recreational sailing, seal viewing excursions, and walking on the mudflats (*wadlopen*), a highly popular activity. The second area is the mainland coastal strip adjacent to the Wadden Sea. Here, tourism remains largely under-developed and agriculture is the dominant economic activity. The third area comprises the barrier islands that protect the Wadden Sea from the North Sea. The five main islands of the Dutch Wadden have developed as popular holiday destinations. In this paper, we specifically focus on the tourism–landscape interactions on the island of Terschelling.

Terschelling comprises a variety of landscape types, including dunes, beaches, forests, meadows, salt marches, and tidal mudflats. The various qualities of these different landscape types make the island very attractive for tourists. Consequently, over the past century, tourism has developed into a well-established industry and currently tourism is the most important source of employment on the island (Sijtsma et al., 2015). In 2017, Terschelling had 4856 inhabitants, about half of whom live in the

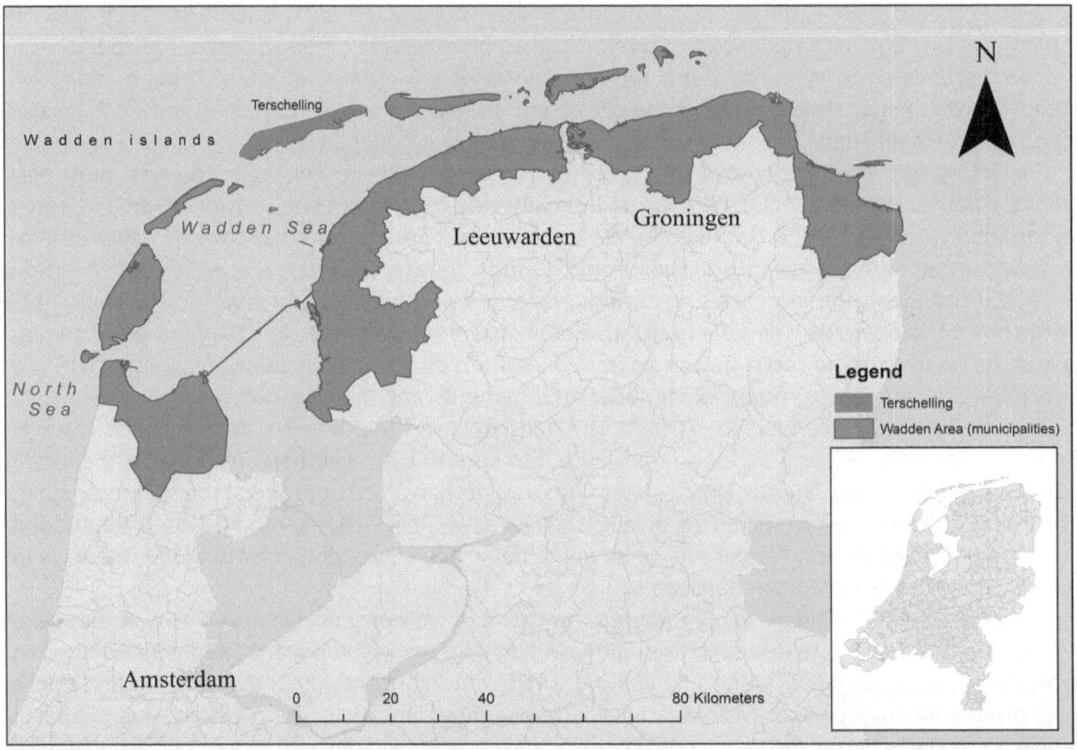

Figure 1. Map of the Dutch Wadden showing the location of Terschelling.

largest town, West-Terschelling, with the remainder spread across nine other villages (CBS Statline, 2017). Terschelling also experiences an annual tourism visitation of around 400,000 tourist arrivals and 1.8 million overnight stays (Municipality of Terschelling, 2014). With a land area of only 86 km^2, there is considerable pressure on its social and ecological carrying capacity. Since the 1950s, there has been a growing demand for accommodation and services to cater for all the tourists. These developments have had negative impacts on the landscape.

Over the last few decades, various national and international laws, regulations and guidelines – including the Bird Directive (1979), Habitat Directive (1992) and Natura 2000 (1992) – have greatly increased the protection of the flora, fauna and ecological state of the area. This increasing regulatory control means that many proposals for socio-economic development have been hindered (Hartman & de Roo, 2013). The twin goals of protecting the island's nature and landscape, and enabling socio-economic development by means of tourism, are heavily debated on the island. There are many stakeholders involved in the governance of T&L at different levels, many of whom have differing and potentially conflicting interests, including tourism entrepreneurs, nature protection organizations, interests groups, governmental bodies and civil society.

Identification and categorization of stakeholders (Results Part 1)

Figure 2 reveals how our panel categorized the identified stakeholders in terms of the interest–influence matrix. It is evident that most tourism entrepreneurs were clustered together in the matrix (see Figure 2 top-left). This observation is supported by what respondents stated in the interviews (e.g. Interviews 9 and 11) – that Terschelling is a relatively small island and that most entrepreneurs are well connected to each other and sometimes transfer business to each other (Interview 9).

Figure 2 also shows that most public institutions form a cluster (across the middle). Within this cluster, the Municipality of Terschelling and the national forest management agency, *Staatsbosbeheer* (SBB), were considered to be the most influential stakeholders on the island, a finding that was confirmed in the interviews (Interviews 1 and 8). Although other clusters could potentially be created, these two clusters (entrepreneurs and public institutions) were seen by the panel members as being important, especially because these two clusters were seen as separate worlds that did not connect with each other. However, a more nuanced picture emerges when the interactions between these two groups are examined closely, which we discuss below.

Interactions between stakeholders (Results Part 2)

To analyze the interactions between the stakeholders, we specifically looked at the three principles of governance we identified from SES theory: inclusiveness, flexibility and multi-scalarity.

Inclusiveness

It was evident from the interviews that many stakeholders were involved in decision-making processes regarding tourism development on Terschelling. Nevertheless, we also observed that there were some stakeholders who were or felt, to some extent, left out. For example, some camping-ground owners on Terschelling thought that their views should be taken into consideration more, and that they were not really involved in decision-making processes, which was frustrating for them (Interview 13). Also, many farmers thought that they were not involved much in these processes, perhaps because they were less interested in tourism development. Nevertheless, some stakeholders considered that there was greater potential for farmers to link with the tourism sector (Interview 11). Instead of intensifying their agricultural businesses, as was happening on the mainland, it was proposed that the island farmers should engage in small-scale organic farming and the selling of local products to tourists (Interviews 11 and 14). The interviewees said this would include the farmers more and would be a better fit with the place-branding characteristics of the island they were trying to promote, which they considered were its small size and its balance with nature.

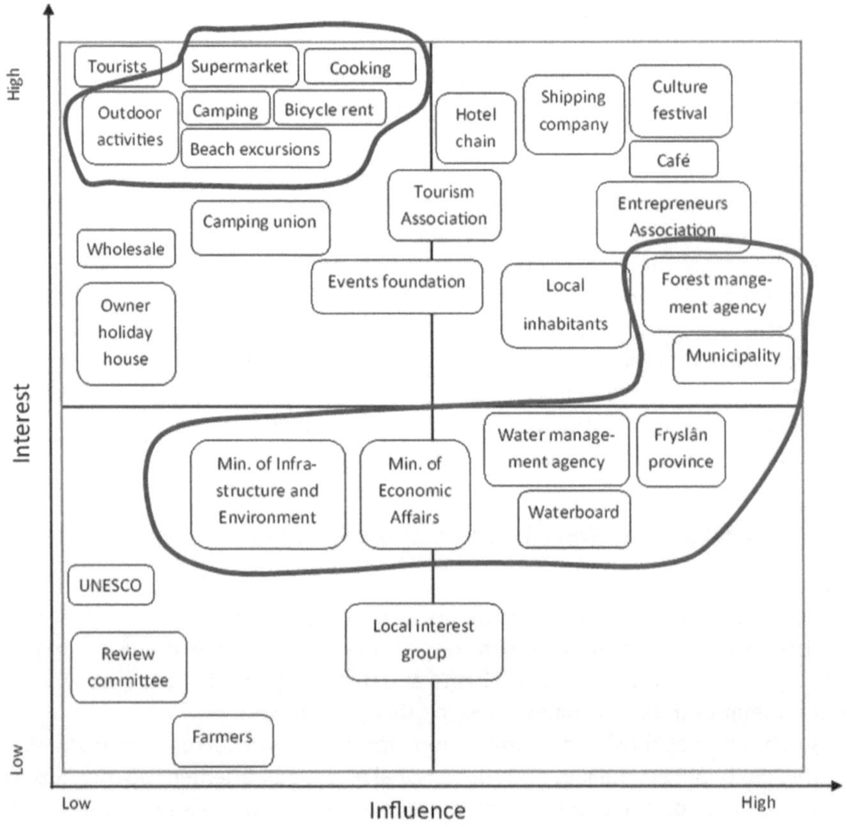

Figure 2. Interest–influence matrix for Terschelling.

To stimulate more inclusion, the municipality has been attempting to get stakeholders involved by organizing discussion groups to consider future developments (Interview 9). However, there was some doubt whether this was tokenistic (and perceived to be a mandatory requirement), rather than being a genuine intention to engage with local people (Interview 10). Also, some interviewees observed that the other influential stakeholder, the forest management agency (SBB), was increasingly including stakeholders in its decision-making processes by sharing and explaining their plans in an open and public way.

> This has resulted in that, almost every year, we have an evening in the pub with SBB, where they explain what they are planning to do. Consequently, they get our support and we appreciate what they do much better. They don't talk about *their* forest anymore, but about *our* forest, and that is how we feel as well. (Interview 11)

Communication is an important component in the effective inclusion of stakeholders (Dare et al., 2014). It was stated that both Key Players on the island (i.e. SBB and the municipality) have recently improved their communication with other stakeholders (Interview 9). While their previous communication practices were considered to be problematic, nowadays SBB was considered to clearly communicate and discuss its intended plans. As a result, they find more support and appreciation for what they are doing.

Flexibility

Developments on Terschelling have been highly regulated since the 1970s (Interviews 2 and 4). However, there were some examples that showed how there can be flexibility as well. In terms of the flexibility of regulation, we observed a difference between those tourism activities having a temporal character and those developments that were more permanent, such as real estate and infrastructure. Activities with a temporal character – such as annual festivals, outdoor activities, seal excursions, and beach excursions – were seen as providing opportunities for development (Interviews 7 and 9). While these activities were considered to be impossible 10–15 years ago (because of the regulatory regime), the interviewees mentioned there have been changes in the management style of SBB. Before these changes, SBB was considered to have had a stubborn attitude, making it almost impossible to negotiate about anything (Interviews 1, 2, 12). One interviewee (an employee of SBB) indicated that SBB nowadays gets into dialogue with tourism entrepreneurs to explore the kinds of activities they want to conduct and to consider under which conditions this could be possible without creating impacts on the landscape: "So, from *'No, unless'* to *'Yes, provided that'*. We [SBB] look at things differently now, which means that we are less likely to wind up in conflict, and we have more real conversations" (Interview 6).

Some interviewees (Interviews 6 and 10) stated that this changing role of SBB can be explained by three factors. First, on a national level, the organization changed its policy to be more "public friendly" and towards one in which communication with stakeholders was considered to be important. Second, on the local level, there were some cultural changes between different generations of local staff of SBB on Terschelling. Third, these changes within the organizational culture of SBB were promulgated by a large budget cut from the national government about 10 years ago. To cope with this cut, SBB had to reinvent itself and had to find other financial sources, which made it much more outward looking. These three reasons led to more dialogue with other stakeholders about possibilities for developing initiatives, instead of SBB telling other stakeholders what was best for the island and what can and cannot be done.

In contrast to these examples relating to temporal activities, there were no examples of flexibility in relation to developments of a permanent character reported to us. This indicates that there are difficulties in the flexibility of the policies. At the local level, in its detailed land-use plans, the Municipality of Terschelling outlines what land uses are permitted on the island and what is not. Camping-ground operators, for example, are entrepreneurs who are subject to these rules and regulations. While these entrepreneurs acknowledged and understood that these rules exist for good reason, they often asked for flexibility to be able to innovate. Some entrepreneurs argued that the tourism market is very fickle, because tourists' preferences are continuously changing, and that to meet changing demands, innovation is continuously required (Interview 4). Camping-ground operators indicated that they wanted to cope with changing demands by implementing quality improvements and improving sustainability aspects of their operations. However, they felt that they were often hindered by the current regulations (Interview 13).

> The municipality should not always try to regulate everything, but they need to cooperate with us and not hinder us. We are already sufficiently constrained by regulation from higher government levels. (Interview 13)

The interviewees stated that the Municipality of Terschelling tends to play it safe when it comes to innovative and creative ideas. Some argued that this was because the municipality lacked a clear vision of future development on the island and they see tourism in only a very simplistic way (Interview 12).

> What worries me is that the municipality puts everything on hold. It does not dare to take responsibility, does not dare to take any risk, and keeps everything out. It is performing a sort of 'village politics', without any vision on where we should go, what we value, and what we should do. (Interview 12)

Respondents argued that the Municipality of Terschelling was incapable of managing the issues on the island. Because the Wadden area is highly regulated at higher government levels, the municipality has a heavy workload, and therefore its staff do not have the time to fully grasp the full meaning of all the policy documents (Interview 10). There was a tendency for the municipality not to take any risks and to leave things as they are, which is likely to hinder Terschelling's full potential into the future. Initiatives that are about the intersection of tourism development and landscape protection may fit with the island's character, but these intersections are difficult for the municipality to deal with, because it is not something that is standardized and consequently may be regarded as too risky (Interview 12). Respondents said that, due to this attitude, initiatives were often postponed for administrative reasons, and that eventually nothing happens. To break through such an impasse, one interviewee suggested that experimentation might be a solution. It was argued that some issues are just too complicated for the Municipality to fully grasp, but simply trying and executing a project can sometimes show the added value of a risky initiative and can help build trust between stakeholders (Interview 12).

Multi-scalarity

We found that many stakeholders on Terschelling were aware that they live in a world that is multi-scalar and that they were dependent on decisions made at higher scales of governance. Many entrepreneurs acknowledged that they "just have to deal with the regulations" from higher levels (Interview 9). However, the way these regulations were implemented by the local government causes dissent. Some stakeholders knew their rights very well and knew how to navigate in the multi-scalar governance system. There was an example of an entrepreneur who disagreed with a decision from the local government. Not satisfied, the entrepreneur pursued his claims and the case ended up at the Council of State, the highest court in the Netherlands (Interviews 9 and 10). In 2010, the entrepreneur won the court case and the municipality had to revise its decision. Interviewees said that, from then on, the municipality's fear of litigation was one reason why the municipality was very cautious in making decisions and was not very willing to support developments they considered to be risky.

For the long-term development of tourism, Terschelling is highly dependent on external actors. Many stakeholders expressed concern about the interference of external actors (i.e. anyone not from the island). For example, external investors, particularly real estate developers, were considered to be a threat because they were considered by some interviewees as having little concern for the local situation and only seeing the island as a site for commercial investment from which they can make a fast profit (Interview 12).

> When you arrive on Terschelling by boat, you think, *'What a charming little harbour!'*. Why should that wonderful feeling be spoilt because some jerk thinks he can construct a high-rise building with lots of apartments and make millions of euros in a very short time and leave us with all the rubbish? (Interview 12)

Some respondents considered another external threat to be interference from mainland consultancy firms. Interviewees said that the municipality has an immense workload and therefore the municipality sometimes hired consulting firms to do various tasks, including social research. It was argued that the mainland firms do not know the local context very well and they often gave the wrong advice. The provincial aesthetics committee was also accused of ignoring and being ignorant of the local context (Interview 11). The aesthetics committee, which is based in Leeuwarden (the capital of the Province of Friesland), comprised architects and other professionals, none of whom were from Terschelling. The committee is charged with assessing building proposals to ensure they meet local aesthetic standards, but in the case of Terschelling, the committee lacked people with appropriate local knowledge. The interviewees said that these external threats can hinder benefit-sharing because, in their view, the benefits accrue to the external investors, while the negative impacts are incurred by the local communities.

Discussion

We believe all three steps in stakeholder analysis are needed to reveal the interactions within and between groups of stakeholders. The first two steps, identification and categorization, helped to show that entrepreneurs are clustered together in the governance structure. Furthermore, despite the fact that they may be competitors, it was shown that they have strong links with each other. In the management literature, this is referred to as "coopetition" (Bengtsson & Kock, 2014). Despite being each other's competitors, the success of each entrepreneur also requires success of the other entrepreneurs. For Terschelling, this means that getting tourists to the island is more important than to which establishment on the island the tourist goes.

The two steps also helped to observe the discrepancy between nature protection policies from public institutions at higher scales and socio-economic developments by tourism entrepreneurs at the local scale, as described by Paloniemi and Tikka (2008). This discrepancy is also considered by Reed et al. (2009), who classified the cluster of entrepreneurs as "Subjects" and the public institutions as "Key Players" and "Context Setters". Subjects, or in our example tourism entrepreneurs, need to establish stronger links and alliances with the public institutions.

The third step, assessing the interactions between stakeholders, is the step we consider most important because it provides complementary qualitative in-depth insights about the interactions. In contrast to the discrepancy described above, this step showed that entrepreneurs and the public institutions were linked quite well, and many stakeholders were involved in decision-making through boards, associations and networks. The three steps of stakeholder analysis, when taken together, help to provide an overview of the stakeholders involved in governance processes and how they interact with each other.

The example of SBB showed that the role of large land owners is crucial in influencing developments in a region. SBB is considered to be a Key Player and how it fulfils its role partially determines the way other stakeholders interact with each other and the benefits that are shared with these stakeholders. The example of the changing role of SBB showed that a large land owner managing a nature area in a stubborn and authoritarian way would be ineffective. Investing in community relations and effective communication, together with an organizational culture change, helped SBB improve the local community's understanding of and support for nature protection and the way they manage the island. This changing role reflects what Eagles (2014) found in the Ontario Provincial Park in Canada. Here, a shift in funding park management from government sources to user-pays operations led to more open-mindedness regarding citizen and client concerns.

We showed that inclusion of stakeholders in decision-making processes is an important condition for benefit-sharing from tourism. Involving all stakeholders is vital for collaboration between stakeholders, because whenever a stakeholder is, or feels, ignored, this could lead to frustration and a lack of willingness to participate. This reflects the idea that collaboration helps avoid conflict (Bramwell & Lane, 2011; Jamal & Stronza, 2009). What was not discussed above was that the inclusion of stakeholders in governance processes is one way to keep the communication lines between different stakeholders open. The case of Terschelling showed that communication between the tourism entrepreneurs and the forest management agency is now perceived to be much better than in the past. Improving collaboration and communication between stakeholders can create mutual understanding and are important preconditions for facilitating benefit-sharing.

Flexibility in decision-making at the local level was considered to be an important factor in facilitating benefit-sharing from tourism in protected areas. Providing flexibility by allowing temporal activities under certain conditions can help explore the possibilities for benefit-sharing in relation to tourism and landscape management. Despite the fact that Terschelling has been highly regulated since the 1970s, flexibility in what was permitted was perceived to be increasing. The interviewees stated that influential stakeholders, such as SBB and the Municipality of Terschelling, played an important role in facilitating this flexibility. Nevertheless, these stakeholders remained hesitant in providing flexibility with regard to developments of a more permanent character. Some interviewees mentioned that this lack of flexibility was problematic, because it made Terschelling vulnerable as a tourism destination in the longer term. Tourism is a sector that changes rapidly, because the demands of tourists are fickle. A tourism destination needs to be flexible and adaptive to be able to cope with changing demands. Wherever a tourism destination is unable to innovate, it runs the risk that tourists will desert it for other places. Not being able to cope with these changes may mean that many (local) stakeholders would miss out on the benefits that might flow from tourism. This makes Terschelling, which is highly dependent on tourism, particularly vulnerable.

We observed that the Municipality of Terschelling struggled to cope, due to the combination of strict and complicated regulations regarding nature protection imposed by the national government (a Context Setter) and because of its limited resources. The interviewees stated that the municipality had difficulties in making clear decisions, tended to avoid risky plans and ideas, and generally "played it safe". This constrained other stakeholders in accomplishing their initiatives and thereby to benefit from tourism. Given these difficulties, the local government is limited in its capacity to fulfil its role as a facilitator of benefit-sharing.

Conclusion

We revealed how the relationships between stakeholders enable and/or constrain the sharing of benefits from tourism in protected areas. The ability to develop synergetic interactions between stakeholders involved in governance processes is likely to be a prerequisite for the effective achievement of benefit-sharing. However, some constraining factors are the lack of capacity of the local government and its ability to cope with future changes. Despite regulations being imposed from above, the way they are interpreted at lower levels is important because this is where most of the opportunities for benefit-sharing lie. We saw that where local government lacks the resources to deal with issues that may foster synergies, it tends to be risk-averse in its decision-making. Consequently, initiatives that could foster benefit-sharing are postponed or obstructed. This means that possibilities for benefit-sharing are often restricted at the lower levels as well as by national and international regulation.

Another constraining factor can be the way an influential stakeholder has interacted with other stakeholders in the past. In our case, the recent positive change in organizational culture within the national forest management agency had implications for the whole island in facilitating benefit-sharing from tourism. How this agency will continue to deal with different stakeholder interests will largely determine the future course of the island. We found that effective communication and an attitude of openness towards all stakeholders were important conditions in facilitating benefit-sharing.

By providing a means to consider the issues faced by all stakeholders, the technique of stakeholder analysis helps to identify the enabling and constraining factors. It helps to systematically provide an overview of who the important stakeholders are, and how they can be categorized. Furthermore, stakeholder analysis provides useful insights into the way stakeholders interact with each other. Using stakeholder analysis as a method also contributes to better understanding governance processes that could help facilitate benefit-sharing from tourism in protected areas.

Governance has recently emerged as a better way of understanding and managing tourism–landscape interactions. We drew on insights from SES theory, in which governance is a key concept. The benefit of using this approach is that SES considers the social and the ecological as an integrated whole, which is helpful for acknowledging synergetic interactions between tourism and landscape. In this paper, we considered three principles: inclusiveness, flexibility and multi-scalarity, and we discussed the way they allow or constrain benefit-sharing between tourism and protected areas. These principles are valuable in providing a deeper understanding of the governance of the interactions between tourism and landscape management.

Based on our research, we have three main recommendations for policy-makers. First, the interest–influence matrix will assist policy-makers to identify and categorize stakeholders. Use of the matrix could be helpful in making strategic choices about how to deal with each type of stakeholder. Second, stakeholder analysis that includes a multitude of different types of stakeholders is a valuable tool to provide an overview that shows which interactions between stakeholders are facilitating and which have limiting effects for benefit-sharing. Third, stakeholder analysis helps policy-makers to intervene more effectively by raising their awareness of the facilitative potential of certain interactions. We hope that with an enhanced understanding of governance processes, policy-makers and landscape managers will be able to increase the potential for benefit-sharing from tourism in protected areas.

Acknowledgments

This work is part of the research programme which is financed by the Province of Fryslân and the Wadden Academy.

Disclosure statement

No potential conflict of interest was reported by the authors.

ORCID

Jasper Heslinga http://orcid.org/0000-0002-6084-9343
Peter Groote http://orcid.org/0000-0002-9642-4013
Frank Vanclay http://orcid.org/0000-0002-9945-6432

References

Adger, W. N., Brown, K., Fairbrass, J., Jordan, A., Paavola, J., Rosendo, S., & Seyfang, G. (2003). Governance for sustainability: Towards a "thick" analysis of environmental decision-making. *Environment and Planning A, 35*(6), 1095–1110. doi:10.1068/a35289

Armitage, D. R., Plummer, R., Berkes, F., Arthur, R. I., Charles, A. T., Davidson-Hunt, I. J., ... Wollenberg, E. K. (2009). Adaptive co-management for social-ecological complexity. *Frontiers in Ecology and the Environment, 7*(2), 95–102. doi:10.1890/070089

Bengtsson, M., & Kock, S. (2014). Industrial marketing management coopetition — Quo vadis? Past accomplishments and future challenges. *Industrial Marketing Management, 43*(2), 180–188. doi:10.1016/j.indmarman.2014.02.015

Berkes, F. (2007). Understanding uncertainty and reducing vulnerability: Lessons from resilience thinking. *Natural Hazards, 41*(2), 283–295. doi:10.1007/s11069-006-9036-7

Bramwell, B., & Lane, B. (2011). Critical research on the governance of tourism and sustainability. *Journal of Sustainable Tourism, 19*(4–5), 411–421. doi:10.1080/09669582.2011.580586

Brondizio, E. S., Ostrom, E., & Young, O. R. (2009). Connectivity and the governance of multilevel social-ecological systems: The role of social capital. *Annual Review of Environment and Resources, 34*(1), 253–278. doi:10.1146/annurev.environ.020708.100707

Buckley, R. (2012). Sustainable tourism: Research and reality. *Annals of Tourism Research, 39*(2), 528–546. doi:10.1016/j.annals.2012.02.003

CBS (2017, May 6). Bevolking; geslacht, leeftijd, burgerlijke staat en regio [Population; gender; age; marital status and region]. Retrieved from http://statline.cbs.nl/StatWeb/publication/?DM=SLNL&PA=03759NED&D1=0&D2=129&D3=1-128,130-797&D4=l&VW=T

Crona, B., & Bodin, O. (2006). What you know is who you know? Communication patterns among resource users as a prerequisite for comanagement. *Ecology and Society, 11*(2), art. 7.

Currie, R. R., Seaton, S., & Wesley, F. (2009). Determining stakeholders for feasibility analysis. *Annals of Tourism Research, 36*(1), 41–63. doi:10.1016/j.annals.2008.10.002

Dare, M., Schirmer, J., & Vanclay, F. (2014). Community engagement and social licence to operate. *Impact Assessment and Project Appraisal, 32*(3), 188–197. doi:10.1080/14615517.2014.927108

Eagles, P. F. J. (2009). Governance of recreation and tourism partnerships in parks and protected areas. *Journal of Sustainable Tourism, 17*(2), 231–248. doi:10.1080/09669580802495725

Eagles, P. F. J. (2014). Fiscal implications of moving to tourism finance for parks: Ontario Provincial Parks. *Managing Leisure, 19*(1), 1–17.

Eagles, P. F. J., McCool, S. F., & Haynes, C. D. (2002). *Sustainable tourism in protected areas: Guidelines for planning and management*. Cambridge: IUCN Publication Unit.

Folke, C., Hahn, T., Olsson, P., & Norberg, J. (2005). Adaptive governance of social-ecological systems. *Annual Review of Environment and Resources, 30*(1), 441–473. doi:10.1146/annurev.energy.30.050504.144511

Foxlee, J. (2007). Key principles and direction for tourism in protected areas: A review of existing charters, guidelines and declarations. In R. Bushell & P. F. J. Eagles (Eds.), *Tourism and protected areas: Benefits beyond boundaries* (pp. 44–69). Wallington: CABI.

Freeman, E. R. (1984). *Strategic management: A stakeholder approach*. Boston, MA: Pitman/Ballinger.

Hartman, S., & de Roo, G. (2013). Towards managing nonlinear regional development trajectories. *Environment and Planning C: Government and Policy, 31*(3), 556–570. doi:10.1068/c11203r

Hartman, S. (2015). Towards adaptive tourism areas? A complexity perspective to examine the conditions for adaptive capacity. *Journal of Sustainable Tourism, 24*(2), 299–314. doi:10.1080/09669582.2015.1062017

Halkier, H. (2014). Innovation and destination governance in Denmark: Tourism, policy networks and spatial development. *European Planning Studies, 22*(8), 1659–1670. doi:10.1080/09654313.2013.784609

Hall, C. M. (2011). A typology of governance and its implications for tourism policy analysis. *Journal of Sustainable Tourism, 19*(4–5), 437–457. doi:10.1080/09669582.2011.570346

Heslinga, J. H., Groote, P. D., & Vanclay, F. (2017). Using a social-ecological systems perspective to understand tourism and landscape interactions in coastal areas. *Journal of Tourism Futures, 3*(1), 23–38. doi:10.1108/JTF-10-2015-0047

Higgins-Desbiolles, F. (2006). More than an "industry": The forgotten power of tourism as a social force. *Tourism Management, 27*, 1192–1208. doi:10.1016/j.tourman.2005.05.020

Higgins-Desbiolles, F. (2011). Death by a thousand cuts: Governance and environmental trade-offs in ecotourism development at Kangaroo Island, South Australia. *Journal of Sustainable Tourism, 19*(4–5), 553–570. doi:10.1080/09669582.2011.560942

Imperiale, A. J., & Vanclay, F. (2016). Experiencing local community resilience in action: Learning from post-disaster communities. *Journal of Rural Studies, 47*, 204–219. doi:10.1016/j.jrurstud.2016.08.002

Jamal, T. (2004). Conflict in natural area destinations: A critique of representation and "interest" in participatory processes. *Tourism Geographies, 6*(3), 352–379. doi:10.1080/1461668042000249656

Jamal, T., & Stronza, A. (2009). Collaboration theory and tourism practice in protected areas: Stakeholders, structuring and sustainability. *Journal of Sustainable Tourism, 17*(2), 169–189. doi:10.1080/09669580802495741

Kabat, P., Bazelmans, J., van Dijk, J., Herman, P. M. J., van Oijen, T., Pejrup, M., ... Wolff, W. J. (2012). The Wadden sea region: Towards a science for sustainable development. *Ocean and Coastal Management, 68*, 4–17. doi:10.1016/j.ocecoaman.2012.05.022

King, B., Pizam, A., & Milman, A. (1993). Social impacts of tourism. *Annals of Tourism Research, 20*(4), 650–665. doi:10.1016/0160-7383(93)90089-L

Kooiman, J. (2003). *Governing as governance*. London: Sage.

Lew, A. A. (2014). Scale, change and resilience in community tourism planning. *Tourism Geographies, 16*(1), 14–22. doi:10.1080/14616688.2013.864325

Libosada, C. M. (2009). Business or leisure? Economic development and resource protection – concepts and practices in sustainable ecotourism. *Ocean and Coastal Management, 52*(7), 390–394. doi:10.1016/j.ocecoaman.2009.04.004

Liu, J., Dietz, T., Carpenter, S. R., Alberti, M., Folke, C., Moran, E., ... Taylor, W. W. (2007). Complexity of coupled human and natural systems. *Science, 317*(5844), 1513–1516. doi:10.1126/science.1144004

Lockwood, M. (2010). Good governance for terrestrial protected areas: A framework, principles and performance outcomes. *Journal of Environmental Management, 91*(3), 754–766. doi:10.1016/j.jenvman.2009.10.005

Luthe, T., & Wyss, R. (2014). Assessing and planning resilience in tourism. *Tourism Management, 44*, 161–163. doi:10.1016/j.tourman.2014.03.011

McCombes, L., Vanclay, F., & Evers, Y. (2015). Putting social impact assessment to the test as a method for implementing responsible tourism practice. *Environmental Impact Assessment Review, 55*, 156–168. doi:10.1016/j.eiar.2015.07.002

McCool, S. F., & Spenceley, A. (2014). Tourism and protected areas: A growing nexus of challenge and opportunity. *Koedoe, 56*(2), 1–2. doi:10.4102/koedoe.v56i2.1221

Mehmood, A., & Parra, C. (2013). Social innovation in an unsustainable world. In F. Moulaert, D. MacCallum, A. Mehmood, & A. Hamdouch (Eds.), *The international handbook on social innovation: Collective action, social learning and transdisciplinary research* (pp. 53–66). Cheltenham: Edward Elgar.

Municipality of Terschelling (2014). *Ontwikkeling aantal overnachtingen op jaarbasis Terschelling* [Development of annual amount of overnight stays on Terschelling] [Data file]. West-Terschelling: Author.

Olsson, P., Folke, C., & Hahn, T. (2004). Social-ecological transformation for ecosystem management: The development of adaptive co-management of a wetland landscape in southern Sweden. *Ecology and Society, 9*(4), art. 2.

Ostrom, E. (2011). Background on the institutional analysis and development. *Policy Studies Journal, 39*(1), 7–27. doi:10.1111/j.1541-0072.2010.00394.x

Paloniemi, R., & Tikka, P. (2008). Ecological and social aspects of biodiversity conservation on private lands. *Environmental Science and Policy, 11*(4), 336–346. doi:10.1016/j.envsci.2007.11.001

Parra, C. (2010). Sustainability and multi-level governance of territories classified as protected areas in France: The Morvan regional park case. *Journal of Environmental Planning and Management, 53*(4), 491–509. doi:10.1080/09640561003737341

Persha, L., Agrawal, A., & Chhatre, A. (2011). Social and ecological synergy. *Science, 331*(6024), 1606. doi:10.1126/science.1199343

Plummer, R., & Armitage, D. (2007). A resilience-based framework for evaluating adaptive co-management: Linking ecology, economics and society in a complex world. *Ecological Economics, 61*(1), 62–74. doi:10.1016/j.ecolecon.2006.09.025

Plummer, R., & Fennell, D. A. (2009). Managing protected areas for sustainable tourism: Prospects for adaptive co-management. *Journal of Sustainable Tourism, 17*(2), 149–168. doi:10.1080/09669580802359301

Prell, C., Hubacek, K., & Reed, M. (2009). Stakeholder analysis and social network analysis in natural resource management. *Society and Natural Resources, 22*(6), 501–518. doi:10.1080/08941920802199202

Prenzel, P., & Vanclay, F. (2014). How social impact assessment can contribute to conflict management. *Environmental Impact Assessment Review, 45*, 30–37. doi:10.1016/j.eiar.2013.11.003

Reed, M. S., Graves, A., Dandy, N., Posthumus, H., Hubacek, K., Morris, J., ... Stringer, L. C. (2009). Who's in and why? A typology of stakeholder analysis methods for natural resource management. *Journal of Environmental Management, 90*(5), 1933–1949. doi:10.1016/j.jenvman.2009.01.001

Revier, H. (2013). Nature conservation and tourism development in the Dutch Wadden Sea region: A common future? In A. Postma, I. Yeoman, & J. Oskam (Eds.), *The future of European tourism* (pp. 177–193). Leeuwarden: European Tourism Futures Institute.

Saarinen, J. (2006). Traditions of sustainability in tourism studies. *Annals of Tourism Research, 33*(4), 1121–1140. doi:10.1016/j.annals.2006.06.007

Sharpley, R., & Ussi, M. (2014). Tourism and governance in Small Island Developing States (SIDS): The case of Zanzibar. *International Journal of Tourism Research, 16*(1), 87–96. doi:10.1002/jtr.1904

Sijtsma, F. J., Broersma, L., Daams, M. N., Hoekstra, H., & Werner, G. (2015). Tourism development in the Dutch Wadden Area: Spatial-temporal characteristics and monitoring needs. *Environmental Management and Sustainable Development, 4*(2), 217–241. doi:10.5296/emsd.v4i2.8561

Sijtsma, F. J., Daams, M. N., Farjon, H., & Buijs, A. E. (2012). Deep feelings around a shallow coast: A spatial analysis of tourism jobs and the attractivity of nature in the Dutch Wadden area. *Ocean and Coastal Management, 68*, 138–148. doi:10.1016/j.ocecoaman.2012.05.018

Snyman, S. (2015). Strategic community participation in sustainable tourism. In S. F. McCool & K. Bosak (Eds.), *Reframing sustainable tourism* (pp. 65–77). New York, NY: Springer. doi:10.1007/978-94-017-7209-9_5

Söderholm, P., & Svahn, N. (2015). Mining, regional development and benefit-sharing in developed countries. *Resources Policy, 45*, 78–91. doi:10.1016/j.resourpol.2015.03.003

Strickland-Munro, J. K., Allison, H. E., & Moore, S. A. (2010). Using resilience concepts to investigate the impacts of protected area tourism on communities. *Annals of Tourism Research, 37*(2), 499–519. doi:10.1016/j.annals.2009.11.001

Swemmer, L., Grant, R., Annecke, W., & Freitag-Ronaldson, S. (2015). Toward more effective benefit sharing in South African national parks. *Society & Natural Resources, 28*(1), 4–20. doi:10.1080/08941920.2014.945055

Vanclay, F. (2012). The potential application of Social Impact Assessment in integrated coastal zone management. *Ocean & Coastal Management, 68*, 149–156. doi:10.1016/j.ocecoaman.2012.05.016

Vanclay, F. (2017). Project induced displacement and resettlement: From impoverishment risks to an opportunity for development? *Impact Assessment & Project Appraisal, 35*(1), 3–21. doi:10.1080/14615517.2017.1278671

Vanclay, F., Baines, J., & Taylor, C. N. (2013). Principles for ethical research involving humans: Ethical professional practice in impact assessment. Part I. *Impact Assessment and Project Appraisal, 31*(4), 243–253. doi:10.1080/14615517.2013.850307

van der Aa, B., Groote, P., & Huigen, P. (2004). World Heritage as NIMBY? The case of the Dutch part of the Wadden Sea. *Current Issues in Tourism, 7*(4–5), 291–302. doi:10.1080/13683500408667986

Waligo, V. M., Clarke, J., & Hawkins, R. (2013). Implementing sustainable tourism: A multi-stakeholder involvement management framework. *Tourism Management, 36*, 342–353. doi:10.1016/j.tourman.2012.10.008

Wang, C. (2012). *A guide for local benefit sharing in hydropower projects*. Washington, DC: The World Bank.

Wesley, A., & Pforr, C. (2010). The governance of coastal tourism: Unravelling the layers of complexity at Smiths Beach, Western Australia. *Journal of Sustainable Tourism, 18*(6), 37–41. doi:10.1080/09669581003721273

Wray, M. (2015). Drivers of change in regional tourism governance: A case analysis of the influence of the New South Wales Government, Australia, 2007–2013. *Journal of Sustainable Tourism, 27*(7), 990–1010. doi:10.1080/09669582.2015.1042482

African tourism industry employees: expenditure patterns and comparisons with other community members

Susan Snyman

ABSTRACT

Tourism is frequently put forward as a means to promote conservation and development. Numerous studies focus on tourists' and tourism industry spending, but very few have looked at tourism staff spending. This paper examines spending patterns of 385 tourism staff in six southern African countries. The analysis includes understanding how much staff are spending, what factors impact on their spending and the local economic impacts of this spending in remote, rural areas. A comparison with 1400 respondents who are not working in tourism (non-staff) highlights the potential indirect multiplier effects and where rural households are spending their income. The results show that tourism staff are spending, on average, more than other community members (non-staff) and that a large percentage of their spending is local and has important positive impacts on other households, highlighting the wider importance of tourism employment, beyond simple job creation and the increased potential benefit-sharing from tourism, through staff spending. Promoting the use of local suppliers of goods and services would increase these induced impacts further. Future research should focus on further rounds of tourism staff spending to determine the full development impact.

1. Introduction

Tourism is frequently put forward as a means to achieve conservation and development goals (Telfer & Sharpley, 2008). 2017 has been declared as the International Year of Sustainable Tourism for Development (United Nations, 2015), further highlighting its global importance and link to the Sustainable Development Goals. However, the capacity of tourism enterprises to reduce poverty in remote, rural areas varies from region to region (Snyman, 2013). Previous research has addressed several aspects of tourism's role in local socio-economic development and poverty reduction through revenue generation (Ahebwa, van der Duim, & Sandbrook, 2012a, 2012b; Novelli & Scarth, 2007), direct employment (Gartner & Cukier, 2012; Mitchell & Ashley, 2010; Snyman, 2014), use of local suppliers of goods and services (Meyer, 2007; Rogerson, 2012), philanthropy (Ashley & Haysom, 2006) and various partnerships with local communities (Ahebwa, van der Duim, & Sandbrook, 2012b; Jones, 2001; Snyman, 2012b, 2014). The basic impact can be augmented by a number of other factors tied to the tourism enterprise's own policies, for example, the managerial style, employment policies and choice of labour intensity, all of which have a direct impact on employment. Exogenous factors can also play a role, either positive or negative; examples include the overall economic situation in the country, population density, and various historical and cultural factors (Spenceley, 2008).

Mitchell and Ashley (2010) describe the direct, secondary and dynamic effects of tourism on the poor. The direct effects include labour income, non-labour income and livelihood effects (Mitchell & Ashley, 2010). Secondary effects include backward linkages and a secondary form of earning called "induced" effects (Mitchell & Ashley, 2010). These induced effects arise when tourism employees spend their earnings locally, thus generating further income for poor households (Mitchell & Ashley, 2010). The focus of this paper will be on these induced effects. It will be shown that these can often be substantial, though they are rarely taken into account, when looking at the local development and benefit-sharing impacts of tourism.

This research is unique in terms of the sample size from across multiple countries in southern Africa. No other paper has examined tourism staff spending across multiple countries. Although Snyman (2012a, 2014, 2017) looked at different aspects of tourism employment, this paper is the first to examine tourism staff spending amounts and patterns in detail across six southern African countries and to examine the impacts of this, providing comparable data and an important analysis of tourism staff spending patterns.

The main focus of the research for this paper was to look at tourism staff spending in six southern African countries and to determine the local socio-economic impacts of this. This was done through conducting extensive socio-economic interview schedules. The paper starts with a literature review of other studies that have looked at staff spending and then describes the methods used in this study and the results found. This is followed by a discussion of the results and conclusions drawn from the research.

2. Literature review

Tourism in this paper is taken to include activities which are nature- and culture-based, sustainable, promote conservation and provide benefits to local people in the area. It is, therefore, not simply tourism that is based on the sale of access to an interesting natural area, but tourism that also provides benefits to local communities. Spenceley (2003) made the point that the long-term sustainability of tourism in Africa will depend on its ability to lift local people out of poverty. This paper looks at the impact of tourism on tourism staff through the payment of wages/salaries and the knock-on effect of these same tourism staff spending these wages/salaries in their home villages.

Sustainability incorporates a balance between economic, social and environmental factors. What is important, particularly in rural areas, is sustainable development as defined by the Brundtland Commission: "development that meets the needs of the present without compromising the ability of future generations to meet their own needs" (World Commission on the Environment and Development [WCED], 1987). While sustainability requires a balance between people, their land-use practices and biodiversity conservation, the particular structure of this balance will differ from country to country, depending on the local socio-economic conditions, political structures and the institutions present (Kepe, Wynberg, & Ellis, 2005). This is important, since the efficacy of conservation efforts on public and private lands in Africa depends, in some measure, on the extent to which these areas are socially, economically and ecologically integrated into the surrounding area (Kepe et al., 2005, p. 8) and the lives of the people in these areas.

Although some writers have attempted to incorporate the relationship between tourism development and poverty alleviation in their research (Ashley, 2000; Ashley, Goodwin, & Roe, 2001; Ashley & Roe, 2002; Ashley & Mitchell, 2005; Mitchell & Ashley, 2010; Reid, 2001; Zhao & Brent Ritchie, 2007), it remains a major shortcoming and needs to be addressed if tourism development is to play a significant role in the alleviation of poverty and related socio-economic development. Goodwin and Santilli (2009, p. 9) also found limited research into the effectiveness of using tourism to deliver economic development and conservation objectives (e.g. Bookbinder, Dinerstein, Rijal, Cauley, & Rajouria, 1998; Lepp, 2007; Shibia, 2010; Walpole & Goodwin, 2001). This paper will illustrate that, to a large extent, tourism is important for socio-economic development in developing countries, but it is not a panacea, and it is important not to exaggerate the benefits or opportunities it can bring, but rather to incorporate it as part of a package of sustainable development options.

There has been a large amount of research on the spending of tourists (Blake, Arbache, Sinclair, & Teles, 2008; Hung, Shang, & Wang, 2013; Mbaiwa, 2005; Mitchell & Ashley, 2010; Muchapondwa & Stage, 2013; Sandbrook, 2010; Scheyvens, 2002; Spenceley, Habyalimana, Tusabe, & Mariza, 2010; Walpole & Goodwin, 2000), but to date no multi-country study has looked specifically at the spending of tourism staff when they are at home and the potential local impacts of this. As tourism wages are mostly re-spent in local villages, they can create new income-earning opportunities for poor people and further increase the benefits of tourism in rural areas. Mbaiwa and Stronza (2010) examined the utilisation of income from tourism in three villages in Botswana and found that the main use of income was buying food, building houses, buying toiletries and clothes, supporting parents, and helping meet expenses associated with school for children. They also found that 8.9% of their respondents saved their tourism-derived income for future uses, such as paying dowry, sponsoring themselves to schools and household emergencies (Mbaiwa & Stronza, 2010, p. 645).

Lapeyre (2011) also examined tourism staff spending in Namibia at one lodge. In Lapeyre's (2011) study, he found that each respondent supported on average four other people, whether financially or in kind. As a result, the lodge delivered benefits for a total of 85 people. Taking the total number of employees working at the Grootberg lodge (23), it was thus estimated that the lodge contributes (both financially and non-financially) to the livelihoods of 115 rural people in the #Khoadi-//Hôas area (23 employees and 92 dependents) (Lapeyre, 2011, p. 231).

In terms of spending, Lapeyre's (2011) study in Namibia found that his respondents were using their wages to build up their physical, financial and human capital. Thanks to tourism revenues, 16 employees (out of 17 interviewed) could further support one or more of their family dependents (Lapeyre, 2011). This change was perceived as the most important livelihood improvement by 41% of Lapeyre's (2011) sample. Lapeyre (2011) also found that most importantly, 64.7% of the employees could use their revenue to cover fees for schools and hostels for their children and their family members' children (Lapeyre, 2011, p. 231). The present study differs from Lapeyre (2011) and Mbaiwa and Stronza (2010) in that it is the first cross-country analysis of tourism staff spending and the sample size is much larger.

Many studies (Bandyopadhyay, Shysamsundar, Wang, & Humavindu, 2004; Barnes, MacGregor, & Weaver, 2001; Mbaiwa, 2003; Muchapondwa, 2003; Muchapondwa & Stage, 2013; Turpie et al., 2006) have looked at the economic impacts of tourism, but the majority have been focused on one country and many have looked only at the impact in terms of the overall income earned from tourism, through GDP, tourism receipts and concession payments. Although Lepper and Goebel (2010) stated that the real significance of ecotourism employment to poverty reduction and livelihood diversification is only fully realised when one considers the "trickle down" of cash income, first to supported family members and then to the greater community, very few studies have looked specifically at the impact of tourism staff spending at the household level. This flow of income from the tourist, through paying for accommodation, curios and activities, to the household level, through the payment of wages and salaries, is critical in terms of understanding the real impact of tourism on poverty reduction and local socio-economic development at the micro-level. Staff spending their salaries in their communities is a direct flow of this income into communities.

Mitchell and Ashley (2010) found that the livelihood effects indirectly caused by the impact of tourist activity on the non-tourist economy can be significant. They also argue that indirect linkages cannot only substantially magnify our understanding of the economic significance of tourism, but can also be particularly effective in transmitting tourism benefits to greater numbers of poor people (Mitchell & Ashley, 2010). The direct employment benefits of tourism are frequently limited to the size of the tourism operation and, therefore, difficult to increase without potential other, possibly negative, impacts especially on the environment. There is, however, potential for increased benefits through indirect and/or induced linkages which may not only be large but also disproportionately pro-poor, compared with direct effects (Mitchell & Ashley, 2010). Mitchell and Ashley (2010) state that there is generally little data specifically on the scale of induced effects. Ardahaey (2011) argues that the magnitude of secondary or induced effects depends on the propensity of businesses and households in the region to purchase goods and services from local suppliers. They go further to state that the induced effects of tourism spending are particularly noticed when a large employer closes an operation (Ardahaey, 2011). Not only are supporting industries (indirect effects) negatively impacted, but the entire local economy suffers due to the reduction in household income within the region (Ardahaey, 2011), highlighting the importance of these impacts. The present study adds significantly to our understanding of the first round of induced impacts of tourism staff spending in rural areas in Africa.

In a study of the impacts of Madikwe Game Reserve in South Africa, Relly (2004, as cited in Rogerson [2006, p. 54]), noted that "the wages earned from the formal lodge industry in a protected area are the single most significant contribution towards poverty alleviation and local economic development and will continue to be so for some time". Similarly, Stronza (2007) and Stronza and Pêgas (2008) stress the overall importance of secure permanent employment that provide a more reliable income stream than other livelihood activities such as agriculture, casual labour or hunting. It is, therefore, important to assess tourism employment's impact on household incomes, staff spending patterns and local economic development in remote rural areas where it is located.

The flow of income from the ecotourist to the household begins with payments for goods and services (e.g. staying in an ecotourism camp or buying a local craft). Some of these payments accrue as direct income to households from goods, others accrue indirectly through the payment of wages and salaries. Some portion of these household incomes then flows to other households as money spent on local products and services. An insight into the details of the process is necessary to any understanding of ecotourism's ability to stimulate local socio-economic development: this is provided in this paper.

There has been much discussion about tourism leakages and linkages (Chirenje, Chitotombe, Gukurume, Chazovachii, & Chitongo, 2013; Mazibuko, 2007; Lange, 2011; Meyer, 2007; Mitchell & Ashley, 2010; Sandbrook, 2010). Sandbrook (2010) argues, however, that where other sources of income are scarce, such as in rural areas in developing countries, tourism can be highly significant in the local context despite considerable leakage. He highlights the importance of the retained income, which although only a small percentage of total income, was still several times larger than income from other sources. Data collected in this study did not provide enough information to calculate average retained incomes across countries, but it would be useful for future research to look at this. Tourism staff spending locally is one approach to increasing local linkages and local multipliers that has not been thoroughly researched. Fennell (2008, p. 81) argues that as new money from tourism enters a local economy it changes hands many times, resulting in a cumulative economic impact that is greater than the initial amount of tourist expenditure. This new money can also come from tourism employees spending their salaries in their home villages and is the focus of this paper. In order to fully understand tourism's local economic development impacts, it is essential that these induced impacts and their multiplier effects are taken into account.

3. Area of study and methods used

In the main study, from which the data in this paper are taken, extensive socio-economic interview schedules were conducted in over 30 rural communities in six southern African countries. The research took place from February 2009 to October 2010. In total, 1785 community interview schedules were conducted in tourism camps (385 staff) and rural villages (1400 non-staff) either within or adjacent to Protected Areas (see Table 1). An attempt was made to interview more than 60% of the community staff at every camp and at least 10% of households in all adjacent communities. Logistically, however, this was not always possible in the non-staff interviews. If the survey area was small, then the majority of the households in the area were interviewed. If the survey area was large, then for logistical reasons, a subset of villages and households was selected. All community staff on duty

Table 1. Study sites and number of interviews.

Country	List of camps surveyed[1]	List of communities surveyed	Number of interview schedules conducted
Botswana	Duba Plains, Vumbura Plains, Little Vumbura	Okavango Community Trust (OCT) villages – Seronga, Gunotsoga, Beetsha, Eretsha, Gudigwa	Staff: 99 Non-staff: 261
Malawi	Mvuu Camp, Mvuu Wilderness Lodge	Balaka District, bordering Liwonde National Park	Staff: 74 Non-staff: 251
Namibia	Skeleton Coast Camp Palmwag Lodge; Doro Nawas Lodge; Damaraland Camp	Okondjombo Conservancy; Purros Conservancy; Sanitatas Conservancy; Orupembe Conservancy Torra, Anabeb and Sesfontein Conservancies	Staff: 81 Non-staff: 271
South Africa	Rocktail Beach Camp Pafuri Camp	Mpukane Community Three villages in the Makuleke community	Staff: 61 Non-staff: 329
Zambia	Kalamu Lagoon Camp	Villages in the Malama Chiefdom adjacent to South Luangwa National Park	Staff: 15 Non-staff: 67
Zimbabwe	Davisons Camp, Makalolo Plains, Little Makalolo and Linkwasha	Villages in Tsholotsho District adjacent to Hwange National Park	Staff: 55 Non-staff: 221

Source: Adapted from Snyman, 2013

and available for interview were interviewed in the tourism camp where they work, while non-staff households were selected randomly, either by walking through the village interviewing every second household (or the next household where someone was available) or interviewing a household member in a public area, e.g. local shop or meeting area. The sample size was not standardised in all study countries due to logistical constraints in some areas where communities were large. These differences in the percentage of the community interviewed could result in some issues relating to external validity. It was, however, felt that all sample sizes were sufficient in the areas surveyed and no new information was found after a certain percentage (approximately 100 households) of the community had been interviewed.

Two types of community member were targeted in the study: those from the community employed in the tourism operation (staff) and those not employed in the tourism operation (non-staff). All respondents lived in, or adjacent to, the conservation area in which the tourism operation was situated. This allowed the comparison of community member's household expenses with those of an equivalent person employed in tourism, in order to provide a comparison to assist in determining the impact of tourism employment and to highlight differences in expenditure patterns between the two groups. Some of the analyses of spending patterns will, however, focus only on the staff sample.

The interviews were conducted by both male and female interviewers, and local translators were used in circumstances where the respondent could not speak or understand English. The interview schedules contained questions relating to demographics, social welfare and living standards, education, employment patterns, income and expenses, health and safety, and attitudes towards tourism and conservation. The majority of questions were quantitative, with some qualitative questions, mostly in the attitudes section of the interview. This paper focuses on the data collected related to expenses. The interview schedule consisted of a structured set of questions, with the majority being close-ended, and some having the option for further explanation. Each interview was conducted verbally, with the interviewer completing the interview schedule. Each interview took approximately 20–45 minutes depending on the respondent's educational level and whether or not translation was required.

Respondents were told that the surveys were confidential and their participation in answering all questions was voluntary. This resulted in some questions not being answered. Non-response to questions did not cluster on particular questions, as no particular question had a greater non-response rate than any other question (Snyman, 2012b). The Zambian sample is included in the analysis because, despite being small, it was felt to be relevant and representative of the area where the interviews were conducted. Inferences drawn from the Zambian results should, however, be viewed with caution.

The interview schedule did not estimate monetary values for subsistence agriculture or use of natural resources, the absence of such values could impact the welfare analysis. This paper, however, focuses specifically on cash income and expenses. As a result of socio-economic differences between countries, the data have largely been analysed by country, rather than aggregated. There are, however, areas of commonality between the research sites due to the remote location of the sites, high levels of poverty and unemployment, and lack of alternative employment opportunities. This commonality allows for some comparison of results, and enables the formation of broad conclusions for the southern African region. A Purchasing Power Parity (PPP) analysis of expenditures is also included to account for country differences. While conducting interviews, the author found that most non-salaried respondents were more aware of their monthly expenditures than they were of their exact monthly incomes (see Snyman, 2013, 2014, for more details on household incomes). This may be explained by respondents' unwillingness to divulge all their income and its sources, while being comfortable revealing all their expenses. While it is unclear from the interview schedules whether or not respondents were spending too much, or were misrepresenting their earnings, it seems likely that the stated monthly expenditures better reflect household "income" than does the respondent's stated monthly household income (Snyman, 2013).

As a result of the diversity in tourism camps, ethnic groups and different land management systems, issues of heterogeneity could be present. As the main aim of the paper was to quantify the impacts of tourism employment per se, it was felt that these differences added qualitatively to the analysis and provided important analyses of comparisons between different areas and countries.

Communities associated with a single tourism enterprise, Wilderness Safaris (WS) were surveyed in this study (for more information on WS, see www.wilderness-safaris.com). It was the only tourism company that had parallel tourism operations, operating according to a standard policy framework in terms of operations and employment policies, over the six Anglophone countries in the region, allowing for greater ease of comparison as these variables could be controlled.

The process followed does, however, mean that some limitations attach to this research. These include that although the camps and communities were diverse (with varying land management systems, ethnic groups, tourism camp price ranges), as only one tourism operator was included in the analysis, there could be limitations to the generalisability of the research. Another is that the author was supported by WS to conduct a study on the impact of the company's activities on rural communities. This study was, however, performed as an independent doctoral researcher looking to discover the realities of tourism and community development and was in no way directly influenced by the company, who merely supported the author logistically and by providing a stipend.

All data collected were analysed using SPSS, v. 12 and STATA v. 10.2, and a combination of descriptive statistics, Probits, linear regressions and Mann–Whitney *U*-tests were used.

4. Results

4.1. Demographic results

In order to contextualise the results analysed in this paper, Table 2 summarises some important demographic statistics from the interviews.

Table 2 shows that staff respondents were, on average, slightly younger and more educated than non-staff respondents. The average household size was found to be similar.

4.2. Dependency levels: number of people supported

Before moving on to a detailed analysis of household expenses, it is important to look at the number of people respondents were supporting, financially or in-kind. The limited formal employment opportunities in remote, rural areas mean that those who do find employment are often supporting a number of people, both direct family as well as others.

Table 3 details the number of people indirectly impacted by the tourism camps in this study, as well as the financial impact of this. The table shows that 16 camps in southern Africa are indirectly

Table 2. Summary of respondents' demographic statistics.

Country	Respondent	Average age (in years)	Gender (% male: %female)	No. of years educated	Average number in the household
Botswana	Staff	31.6	46:5	8.7	4.1
	Non-staff	33.5	43:6	6.6	6.6
Malawi	Staff	36.3	88:1	8.1	5.5=
	Non-staff	38.7	39:6	3.2	4.6
Namibia	Staff	32.5	40:6	9.8	8 .6=
	Non-staff	36.8	55:5	5.1	8 .2=
South Africa	Staff	36.7	39:6	10.3	6 .9=
	Non-staff	44.3	31:7	6.9	6.4
Zambia	Staff	36.8	100:0	8.9	6.4
	Non-staff	40.9	27:7	5.7	5.9
Zimbabwe	Staff	36.2	100:0	9.1	7.1
	Non-staff	43.5	35:6	6.9	6.7
Average	Staff	34.4	61:4	9.1	6.3
	Non-staff	39.5	40:6	5.8	6.5

Source: Snyman (2013).

impacting nearly 5000 people's lives; equivalent to 14 people per tourism bed (at 100% occupancy). This result is, however, affected by the findings in Botswana (27 people/bed). The result is high in Botswana, largely due to the high cost of the tourism product at the Botswana study sites and, therefore, high number of staff employed. Fourteen people per tourism bed are, however, substantial considering the rural nature of the areas where the camps are situated, the lack of permanent employment

Table 3. Number of people indirectly affected by tourism employment.

National group sampled	Total no. of staff in the surveyed camps	Average stated monthly wage per staff member (USD 2011)[1]	Average no. of dependents per staff respondent	Total no. of people indirectly impacted by camp employment[2]	Average stated monthly amount given to dependents per staff respondent (USD 2011)[3]	Total payments to dependents per month (USD 2011)[4]	Approximate number of people indirectly impacted per tourism bed
Botswana (3 camps surveyed – 52 beds)	173	220.01	8	1384	$39.03	$6752.19	27
Malawi (2 camps surveyed – 50 beds)	108	83.73	8	864	$5.78	$624.24	17
Namibia (4 camps surveyed – 104 beds)	166	172.65	6	996	$31.64	$5252.24	10
South Africa (2 camps – 64 beds)	94	319.94	6	564	$20.21	$1899.74	9
Zambia (1 camp – 18 beds)	23	159.40	7	161	$6.15	$141.45	9
Zimbabwe (4 camps – 64 beds)	119	216.87	8	952	$11.30	$1344.7	14
Average/Total (16 camps – 352 beds)	683	USD 195.43	7	4921	USD 19.02	USD 12,990.66	14

1 Over and above wages employees receive gratuities (not included in this analysis) as well as other non-monetary benefits of employment such as accommodation, food, uniform, and a company HIV awareness/testing and education programme. These figures are based on data from the socio-economic interview schedules and are not official wage figures; they are based on the salary figures given by respondents. These figures have been inflated to 2011 values.

2 This result is calculated by multiplying the number of people employed in the surveyed camps by the calculated average number of staff dependents. All figures are rounded up.

3 These figures were obtained from the expenses section of the interview schedules conducted in the study countries and exclude money given for education, living expenses, etc. This amount is a remittance amount.

4 These figures were calculated by multiplying the total number of staff by the total monthly payment to dependents.

and high levels of poverty. Employees of the camps were found to be averaging total transfers of USD19 per month to dependents; almost USD13,000 paid monthly. This excludes money given for education, food and other living expenses. In terms of the number of jobs per camp, the figures in Table 3 give an approximate number of 43 direct jobs per tourism camp studied (this will obviously depend on the camp size) or 2 direct jobs per tourist bed (at 100% occupancy). Policy implications of this data include that it can be used in future tourism developments in rural areas, by assisting tourism operators in providing the average figures for tender documents and for determining the potential impact their operations may have in rural areas, providing concrete data for potential socio-economic returns for government to assess.

Overall, there was a significant statistical difference found between the mean number of dependents for all staff ($M = 7.3$) and all non-staff respondents ($M = 4.9$) ($U = 151,382$, $p < .001$). These high dependency ratios indicate the breadth of tourism's indirect impact among local communities in these areas. Table 3 also illustrates that money given to dependents was not proportional to wage/salary, but may have been more dependent on needs to provide money for other expenses, e.g. living expenses, education, etc. For example, in Zimbabwe, the cost of living is, generally, higher than in Botswana, and there is a cost to education, which is free in Botswana.

4.3. Analysis of factors impacting spending

Table 4 presents the regression results for monthly household expenses (to be discussed in detail in the next section). The table shows an analysis of each country separately, the full sample and also for the non-staff sample only. The number of dependents was significant in all countries, except Botswana and Zambia, with a positive relationship to household expenses, i.e. as the number of dependents increased, the household expenses tended to increase. Those with higher education levels, more income sources, family employed in tourism/conservation and employed by WS (except in Malawi) tended to have higher monthly household expenses. In general, ceteris paribus, younger respondents (except in Botswana) tended to spend less than older respondents, though not significantly. Households with a male household head tended to spend more (except in Namibia) and those currently employed (in any formal job) also tended to have higher expenses.

One of the key arguments in support of tourism is that it injects cash into rural areas, which can then help kick-start other enterprises, via a local multiplier effect. In order to evaluate the relevance of this argument, this study analysed the spending of staff and non-staff respondents. Analysis did not, however, extend beyond the respondents' first round of spending, which is a limitation of this study. Local economy is defined in this study as the villages where staff live and are spending the majority of their wages/salaries. In most cases, these are remote, rural villages adjacent to, or in, conservation areas. Although goods may be imported to the village shops from larger economic centres, this paper is only looking at the first round of spending and the impacts of this.

Table 5 shows monthly household expenditures for staff and non-staff respondents in each country, including a PPP comparison, illustrating a statistical difference between staff and non-staff respondent expenditures in all countries.

In order to assess the first round of possible local linkages, Table 6 presents the mean monthly household expenditures of staff respondents only and projected expenditures calculated for the entire workforce of the surveyed camps.

In Table 6, it shows that, on average, the interviewed tourism staff were spending approximately USD278 each per month, with South African staff spending the most per month and Malawian staff the least. Calculations in Table 6, based on the mean monthly household expenditures reported by staff and the total number of staff employed in the surveyed camps, suggest that staff inject a total of USD212,144 monthly into local economies where they live across the six countries. Some of this accrued from other sources, but the bulk was from tourism (see Snyman, 2014). This yields approximately USD13,259 per month per tourism camp or USD603 per tourism bed per month, with Botswana having the highest tourism staff spend per bed and Malawi the lowest.

Table 4. Expense regression results by location.

Variable	Botswana (n = 311)	Malawi (n = 294)	Namibia (n = 294)	South Africa (n = 351)	Zambia (n = 80)	Zimbabwe (n = 259)	Full sample (n = 1513)	Non-staff sample (n = 1248)
Intercept	.328	2.25***	4.93***	4.07***	2.25***	2.93***	3.85***	3.28***
	(.542)	(.325)	(.240)	(.414)	(.325)	(.466)	(.190)	(.210)
Country 1: Botswana	N/A	N/A	N/A	N/A	N/A	N/A	−1.06***	−1.41***
							(.101)	(.120)
Country 2: Malawi	N/A	N/A	N/A	N/A	N/A	N/A	−1.40***	−1.48***
							(.120)	(.132)
Country 3: Namibia	N/A	N/A	N/A	N/A	N/A	N/A	.187*	.208*
							(.102)	(.122)
Country 5: Zambia	N/A	N/A	N/A	N/A	N/A	N/A	−.905***	−.918***
							(.150)	(.175)
Country 6: Zimbabwe	N/A	N/A	N/A	N/A	N/A	N/A	−.548***	−.648**
							(.105)	(.124)
Environmental income	.732	.218	NI	−1.45**	.192	.197	.057	.113
	(.538)	(.275)		(.657)	(.935)	(.488)	(.216)	(.226)
Formal employment	.675**	−.037	−.228	−.354	.316	.076	.112	.106
income	(.269)	(.160)	(.177)	(.283)	(.315)	(.215)	(.088)	(.094)
Piecework/casual	.576*	−.501*	−.093	−.786**	−.185	.109	−.127	−.078
labour	(.298)	(.271)	(.303)	(.329)	(.353)	(.210)	(.113)	(.119)
Government grants	.632	NI	−.377	−.506*	NI	−.295	.226*	.141
	(.397)		(.213)	(.280)		(.430)	(.115)	(.125)
Tourism-related	.791*	1.26*	−.134	−.820**	NI	.062	.242	.222
income	(.453)	(.709)	(.205)	(.396)		(.288)	(.152)	(.190)
WS employed	1.93***	−.645	.587**	1.01**	.301	.886**	.945***	NI
	(.442)	(.709)	(.229)	(.321)	(.428)	(.323)	(.156)	
Age	.012	−.005	−.003	−.002	−.011	−.006	−.003	−.006*
	(.011)	(.010)	(.006)	(.005)	(.048)	(.007)	(.003)	(.003)
Male	.047	−.164	.261**	.067	−.046	−.045	.096	.101
	(.195)	(.154)	(.124)	(.107)	(.451)	(.187)	(.068)	(.079)
No. of children	.001	.019	.026	.050*	−.129*	−.004	.014	.017
	(.052)	(.029)	(.021)	(.028)	(.065)	(.033)	(.014)	(.015)
No. of dependents	.019	.070**	.019**	.034*	.054	.062**	.037***	.034***
	(.021)	(.021)	(.007)	(.019)	(.045)	(.024)	(.007)	(.007)
No. in the household	−.027	.080	.001	.009	.316**	.024	−.0001	.0009
	(.031)	(.055)	(.010)	(.025)	(.134)	(.027)	(.0080	(.009)
No. in household under	.061	−.110*	.011	.009	−.257*	−.032	−.006	−.002
20 years old	(.040)	(.061)	(.016)	(.036)	(.139)	(.029)	(.013)	(.014)
Male household head	.362*	.468**	−.097	.236**	.450	.300*	.308***	.348***
	(.191)	(.169)	(.126)	(.104)	(.407)	(.175)	(.069)	(.079)
Age of the household	−.004	−.001	−.0008	.007*	.001	−.001	−.004	−.0007
head	(.006)	(.009)	(.004)	(.004)	(.047)	(.006)	(.002)	(.003)
No. of years educated	.112***	.060**	.009	.027*	.040	.073**	.035***	.034**
	(.030)	(.018)	(.017)	(.016)	(.046)	(.027)	(.009)	(.010)
Currently formally	.846**	.520	.459**	.312*	.655	.316	.539***	.522***
employed	(.283)	(.287)	(.171)	(.179)	(.567)	(.397)	(.111)	(.117)
No. of income sources	.863***	.257**	.158**	.177**	.275*	.249**	.322***	.353***
	(.139)	(.094)	(.073)	(.057)	(.147)	(.077)	(.038)	(.043)
Family employed in	.154	NI	−.204*	.061	−.181	.467**	.071	.119*
tourism/conservation	(.163)		(.117)	(.096)	(.252)	(.135)	(.060)	(.070)
Adjusted R^2	0.4741	0.3470	0.1333	0.2484	0.4784	0.3261	0.4317	0.3924
Prob > F	0.0000	0.0000	0.0000	0.0000	0.0000	0.0000	0.0000	0.0000

Notes: The figures in parentheses are standard errors. ***, ** and * indicate significance at the 1%, 5% and 10% levels, respectively.

Not included (NI): There were no government grants in Malawi and family employed was not included in the staff interview schedules: these variables were, therefore, not included in the regression.

In Namibia, no environmental income was reported.

In Zambia, there were no government grants and tourism-related income correlated with WS Employed, and was therefore excluded from the regression.

Environmental income refers to any income derived from the environment, e.g. selling wood, selling herbs/plants from the forest, etc.

Table 5. Monthly household expenditures by location.

National group sampled	Sample size (n)	Mean monthly household expenditures (USD 2011)	Log monthly expenditures	Statistical significance	Purchasing Power Parity conversion[1] for mean stated monthly household expenditures (USD 2011)
Botswana – staff	99	319.72	$M = 5.60$, SE = .06	$U = 2697.50, Z = -11.603,$	547.86
Botswana – non-staff	261	112.46	$M = 3.36$, SE = .12	$p < .001$	190.61
Malawi – staff	74	113.87	$M = 4.41$, SE = .08	$U = 4016.0, Z = -7.42,$	284.68
Malawi – non-staff	251	50.57	$M = 3.33$, SE = .08	$p = .001$	126.43
Namibia – staff	81	317.41	$M = 5.12$, SE = .06	$U = 8126.5, Z = -3.545,$	406.94
Namibia – non-staff	271	258.35	$M = 5.12$, SE = .06	$p < .001$	331.22
South Africa – staff	61	400.37	$M = 5.85$, SE = .06	$U = 3165.0, Z = -8.495,$	548.45
South Africa – non-staff	329	209.09	$M = 4.96$, SE = .06	$p < .001$	286.42
Zambia – staff	15	151.51	$M = 4.91$, SE = .12	$U = 251.0, Z = -3.017,$	172.17
Zambia – non-staff	67	99.70	$M = 3.93$, SE = .17	$p < .05$	113.29
Zimbabwe – staff	55	367.44	$M = 5.60$, SE = .09	$U = 1946.0, Z = -7.80,$	Not available[2]
Zimbabwe – non-staff	221	135.53	$M = 4.38$, SE = .08	$p < .001$	Not available[2]

[1] PPP conversion factors were obtained from www.tradingeconomics.com; these figures were taken from a 2012 World Bank report.
[2] PPP conversion factors were not available for Zimbabwe.

Since there are few commercial shops in the villages close to the camps, some household expenditures are made directly to other households in the local economy surrounding the camps. On the other hand, some will be spent on transport and on goods brought in from larger centres. In each country, the flow of tourism income into the local economy is an injection of cash that can then be circulated to stimulate further spending and, therefore, increase the local multipliers and linkages.

Below are two examples of the benefits received by two tourism employees in the study and the associated local knock-on effects of employment. The first example is of a 37-year-old female employee in Namibia from a village within the Torra Conservancy who started working as a housekeeper and then moved to waitress, bar lady, supervisor, assistant manager and is now camp manager. Over and above her monthly wage, she receives food, accommodation, transport, pension, uniform and medical aid, and received training from WS in service, housekeeping, computers, driving, HIV and management. She supports six people and her children attend the local school. Consumer durables in her household included a television, fridge and mobile phone. She estimated that 30% of

Table 6. Staff sample: total monthly cash injection from tourism into the local village economy.

National group sampled	Mean monthly household expenditures – staff only (USD 2011)	Total no. of staff in surveyed camp	Projected total monthly injection into the local village economy (USD 2011)[1]	Projected mean cash injection into local economy per tourism bed per month (USD 2011)
Botswana (3 camps surveyed – 52 beds)	319.72	173	55,311.56	1063.68
Malawi (2 camps surveyed – 50 beds)	113.87	108	12,297.96	245.96
Namibia (4 camps surveyed – 104 beds)	317.41	166	59,690.06	573.94
South Africa (2 camps – 64 beds)	400.37	94	37,634.78	588.04
Zambia (1 camp – 18 beds)	151.51	23	3484.73	193.59
Zimbabwe (4 camps – 64 beds)	367.44	119	43,725.36	683.21
Average (16 ecotourism camps – 352 beds)	278.39	683	212,144.45	602.68

[1]Calculated by multiplying staff respondents' mean stated monthly household expenditures by the number of staff at the surveyed camp.

her monthly expenditure is directed to the local shops and spent on airtime, food and cleaning materials. The second example is of a 35-year-old male employee in Malawi who started working as a trainee waiter, moving on to barman and is now a safari guide. He received training in food and beverage and safari guiding from WS. Over and above his monthly wage, he receives food, pension and medication. He comes from one of the villages adjacent to the Park and supports approximately 15 people. Consumer durables in his household included a television, fridge, two mobile phones and a portable radio. The majority of his expenditures were in the local economy and were on school fees, maize, cassava and other food goods bought from a local store. These examples illustrate the role of tourism employment in reducing poverty by increasing real household incomes and opportunities, through staff's ability to invest in assets and educating their children.

4.4. Spending patterns of tourism staff and other community members

In order to further understand the impact of tourism on rural communities, this section details tourism staff spending patterns.

Table 7 details the reported expenditures of staff and non-staff respondents on major items, including the US dollar equivalent and the percentage of total expenses. This table is aggregated for all respondents. In total, staff were spending 28% of their monthly household income on food, while non-staff respondents were spending nearly 36%. As the majority of staff respondents receive meals when on duty, spending on food is largely for their dependents.

Overall, staff expenditures were high on the following items: education, money given to dependents, accounts/lay-byes, casual labour, telephone, toiletries and water expenses.

Non-staff expenditures were high on toiletries, cleaning materials, transport, clothes, education and loans. Overall, they were, in general, spending more than staff respondents on necessities, such as food, toiletries, transport and gas.

The expenditure on food in this analysis was for "purchased foods" only, and therefore excludes the value of all subsistence farming and bartered foodstuffs. While conducting interviews, it was observed that most households engaged in some home cultivation, particularly maize, pumpkin, millet, groundnuts, watermelon and sorghum. Some households had larger plots and were able to sell surplus produce.

Table 7 shows high non-staff payments towards loans in Malawi, Botswana and Zambia. Long distances between villages in Namibia explain the high expenditure on transport. The casual labour expenditure in Botswana was largely for cattle herders who live at cattle posts some distance from the village. A number of households in Malawi rely on annual cash crops (cotton, tobacco) resulting in the need for credit before the markets open each year.

5. Discussion

An important aspect of tourism's local economic development potential is whether or not staff wages/salaries are being spent locally or the money is leaking out to larger cities, or in some cases, other countries. It was observed in this study that tourism staff expenditures on casual labour, money to dependents, gas/paraffin/candles, alcohol and rent were frequently made into the local economy, setting up important local community knock-on/multiplier effects. Leakages out of the local economy included expenditures on telephone airtime, some clothes and toiletries, although these were still frequently made at a local shop, who bought the goods in from a larger centre, and, therefore, still resulting in an impact in the local village economy. Expenditures at the clinic (medical), education, water, electricity are largely to local government, and therefore important in terms of local economic development. Numerous small businesses (bakeries, grocery shops, tailors) were observed in the villages: tourism staff's ability to spend at these businesses stimulates cash flow in the local economy and, therefore, further rounds of spending. As was found in the present study, in Bryceson (1999), many case studies from the De-Agrarianisation and Rural Employment (DARE) programme (Nigeria,

Table 7. Staff and non-staff monthly household expenditures, as a percentage and as a USD amount.

Staff (*n* = 385)	% of total expenditures	Total USD amount	Non-staff (*n* = 1400)	% of total expenditures	Total USD amount
Food	28.0%	35,716.07	Food	35.5%	78,567.99
Education	11.3%	14,348.65	Toiletries	6.7%	14,841.64
Money to dependents	7.2%	9215.10	Cleaning materials	6.2%	13,652.04
Telephone	5.6%	7151.76	Transport	5.8%	12,736.57
Accounts	5.6%	7073.94	Clothes	5.2%	11,474.64
Toiletries	5.1%	6441.51	Education	5.0%	11,043.21
Clothes	4.7%	5959.20	Loans	4.9%	10,946.63
Casual labour/ childcare	4.6%	5872.97	Alcohol	4.2%	9276.13
Other	4.6%	5859.24	Other	3.9%	8529.31
Rent	3.9%	4960.00	Money to dependents	3.7%	8152.04
Cleaning materials	3.6%	4634.70	Casual labour	3.0%	6610.80
Electricity	3.0%	3859.74	Gas/paraffin/ candles	2.9%	6427.07
Medical Expenses	2.6%	3276.90	Telephone	2.8%	6297.46
Transport	2.1%	2638.01	Accounts	2.8%	6187.59
Gas/paraffin/candles	1.8%	2290.55	Electricity	2.2%	4807.02
Loans	1.8%	2236.37	Cigarettes	1.9%	4262.39
Water expenses	1.6%	2055.09	Medical expenses	1.6%	3440.37
Alcohol	1.4%	1847.58	Fuel	1.5%	3252.15
Fuel	0.8%	1030.52	Rent	0.3%	656.54
Cigarette	0.8%	972.28	Water expenses	0.1%	238.55
Total	100.0%	127,440.19	Total	100.0%	221,400.15

Ethiopia, Zimbabwe) also found that the largest expenditure was on staple food purchases. The same has been found in country Household Income and Expenditure Surveys (Namibia, Botswana). In Lepp's (2007, p. 881) study in Uganda, "an elderly man, still farming, explained: 'when those people get money from tourism so do we because they buy food from us' (Mzee Isabirye, 2003, personal communication)". Similarly, many non-staff respondents in this study mentioned that tourism staff spending in the villages benefitted them and their households (Snyman, 2014). This highlights the role of tourism as a vehicle for local economic development, community development, poverty reduction and sustainability. Criticism related to the limitation of growth in terms of direct tourism employment can be somewhat offset by the broader real impacts of tourism through existing tourism employee spending in a local economy. This first round of spending, which can stimulate further rounds of spending (beyond the scope of this paper), can have a real development impact and should not be underestimated.

Stem, Lassorie, Lee, Deshler, and Schelhas (2003) and Stronza (2007) found in their case studies that people employed in tourism spent their "new" income on household needs or "family well-being" – basic subsistence, health, education and home improvements. Overall, the present study found similar results in terms of household expenditure patterns, with tourism staff members spending a large majority of their household income on education and, therefore, building human capital. This also plays an important role in terms of the real development impacts of tourism in developing countries.

In terms of the dependency levels, a high number of dependents tend to erode savings, and although tourism staff should be able to accumulate wealth, they are often unable to do so because they support a greater number of people. The Managing Director of the Malawi operation discussed this issue with the author and introduced a pension system for staff to institutionalise saving (Snyman, 2013). In the author's experience, however, the pension deduction is frequently seen negatively as a loss of immediate cash income, rather than as a saving and was frequently mentioned as an expense by many staff interviewees. Education related to this is important in terms of long-term sustainability and to promote long-term saving to make a real difference in reducing household poverty. For tourism and the related employment to truly impact positively on development, it has to have real long-term impacts on poverty and this often requires a longer term perspective rather than immediate returns from a salary or income provided to a community through a lease or revenue share. This can occur through pension programmes, investment in assets, human capital, etc.

The ability of staff to spend on "luxury" items such as rent, telephone and water affords them the ability to improve their overall social welfare. The high expenditure by staff on accounts/lay-byes needs to be assessed further and financial advice should be given to staff in terms of the pitfalls of buying on credit. As mentioned above, spending on children's education is, however, an investment in human capital and is consequently an important household diversification strategy.

Tourism employment was shown in this paper to have an impact on overall total household welfare through the provision of salaries/wages to support household spending on both necessities and luxury goods, highlighting tourism's important role in local economic development and long-term poverty reduction. The ability of tourism staff, through their regular, reliable monthly income, to invest in assets can have a significant impact on the long-term welfare of households and on their ability to cope with shocks, risks and other economic stresses. This then extends tourism impacts from a micro- to a more macro-level, having impacts beyond the actual tourism business itself. The flow of tourism income into the communities through staff spending their salaries in their home villages further extends the impact of tourism on poverty reduction and social welfare to households other than those who have a family member employed in tourism (a detailed analysis of this was, however, beyond the scope of this study).

Ways to improve and ensure the sustainability of the induced impacts of those employed as well as other households in the area include ensuring that tourism operations are paying equitable wages and salaries to employees and employing across a broad spectrum of local families rather than from one family, thus spreading benefits further; providing financial training and courses to communities/tourism staff in order to ensure that communal funds, as well as individual salaries/dividends, are properly managed and that there are provisions for the future through savings, pension funds, education funds, etc. An important part of this would include providing financial advice to staff in terms of the potential pitfalls of buying on credit; encouraging (and possibly, where appropriate, even stipulating in tourism concession agreements) tourism operations to use local suppliers of goods and services to further enhance local linkages and local multipliers; in conjunction with the latter, local suppliers may need guidance if they are to provide goods and services of the correct quality and quantity for tourism operations, invest money back into their businesses, and plan for the future; and encouraging entrepreneurs to set up small businesses in villages to supply goods and services to local villagers as well as the tourism industry e.g. fresh produce markets, dry goods shops, crafts, etc. This will result in further rounds of local spending and benefits.

Many of the above are ostensibly functions of government, but it was clear that in a number of the six countries researched, the private sector and NGOs will have to fulfil this role or at least provide support to government in these roles.

6. Conclusion

If markets are functioning efficiently and effectively they allow opportunities for trade and exchange which, in turn, increases the circulation of cash in rural areas and gives the poor greater opportunities to find ways to lift themselves out of poverty (Ellis, Kutengule, & Nyasulu, 2003; Ellis & Mdoe, 2003; Freeman, Ellis, & Allison, 2004). Evidence of this was seen in the study areas where small shops had been established to offer goods and services to employed individuals (largely tourism staff in these areas) in the area. Local villagers also mentioned the benefit of employment in the fields or households of employed tourism staff: a further development impact of tourism in rural areas. The results of this paper highlight that there is an important avenue for tourism to increase local multipliers through tourism staff spending locally in their home villages (induced impacts of tourism). As the analysis in this paper was only on the first round of tourism staff spending, there is a need to research induced tourism impacts and the associated multiplier effects further in order to fully understand the impact that tourism staff spending can and does have on the local economy and local community development and poverty reduction.

It has been shown here that tourism employees' ability to invest in human capital (through providing education for family members) and purchasing various assets provides opportunities to cope with future vulnerability and the adoption of more efficient livelihood strategies (Lapeyre, 2011, p. 231), illustrating broader real development impacts of tourism. Investing in the education of one's children can be seen as a long-term livelihood diversification strategy, being an investment in the future resource base of the household, with the hope of future income returning to the parental household (Ellis, 1998; Niehof, 2004). Tourism employment, therefore, has the potential to assist in long-term stability for households, reducing risk and vulnerability, and therefore the stress faced by households. This can, however, lead to a deepening of inequalities in rural areas as staff are able to reduce risks, earn more income and educate their families further, while other households remain "trapped" in a subsistence lifestyle, vulnerable to shocks (Snyman, 2014). The community way of living does negate this somewhat as staff support their extended families and, even in some cases, friends. Ogutu (2002) similarly found that the accumulation of savings by individuals in his Kenyan study was leading to social differentiation that went beyond traditional realms. This is an important factor that needs to be borne in mind when assessing the "benefits" of tourism in rural communities as shifts/changes in traditional community structures can lead to a number of other, unanticipated effects that may, or may not, be positive. It is important though, in the long term, that there is a broadening of the benefits and encouragement, empowerment and education of rural households to diversify their livelihoods and to assist them in finding ways out of poverty; tourism being only one of a suite of options. Improvements in local institutions and governance structures in rural areas will need to form a large part of this. This possible social differentiation is important from a development perspective and further research on it is encouraged.

The implication from the results in this paper is that tourism has a local knock-on effect because much of the budget of tourism camps is spent on wages and much of these were found to be spent locally in their home villages. Staff's higher expenditure in areas such as education and supporting dependents highlights tourism employment's contribution to local social welfare and poverty reduction, as well as improvements to human capital in remote, rural areas. It was also shown that tourism staff are providing support to a large number of dependents both through giving cash directly to dependents, as well as providing food, clothes and education to them.

Simply looking at tourism camps' direct employment figures, therefore, significantly underestimates the development impacts of tourism in rural areas. Illustrated in this paper is the fact that other than tourism itself bringing immediate and important direct economic and financial benefits to people in rural areas, specifically the staff employed directly in tourism operations, it also has the potential to provide benefits to other households when tourism staff spend their salaries in rural areas. These induced impacts and the associated multiplier effects should be researched further and in greater detail in order to fully understand the potential local development and poverty reduction impacts of tourism.

Acknowledgments

The author would like to gratefully acknowledge funding from the Swedish International Development Cooperation Agency (SIDA) through the Environmental Economics Policy Research Unit (EPRU) at the University of Cape Town. Special thanks also to Wilderness Safaris for accommodation, transport and logistical support, and to the journal editors and anonymous reviewers who provided valuable feedback and comments.

Disclosure statement

The author was supported by Wilderness Safaris to conduct a study on the impact of the company's activities on rural communities. This study was, however, performed as an independent doctoral researcher looking to discover the realities of tourism and community development and was in no way directly influenced by the company, who merely supported the author logistically and by providing a stipend.

Funding

The author would like to gratefully acknowledge funding from the Swedish International Development Cooperation Agency (SIDA) through the Environmental Economics Policy Research Unit (EPRU) at the University of Cape Town.

References

Ahebwa, W. M., van der Duim, R., & Sandbrook, C. (2012a). Tourism revenue sharing policy at Bwindi impenetrable national park, Uganda: A policy arrangements approach. *Journal of Sustainable Tourism, 20*(3), 377–394.

Ahebwa, W. M., van der Duim, R., & Sandbrook, C. G. (2012b). Private-community partnerships: Investigating a new approach to conservation and development in Uganda. *Conservation & Society, 10*(4), 305–317.

Ardahaey, F. T. (2011). Economic impacts of tourism industry. *International Journal of Business and Management, 6*(8), 206–215.

Ashley, C. (2000). *The impacts of tourism on rural livelihoods: Namibia's experience* (Working Paper 128). Overseas Development Institute. Retrieved December, 18, 2008 from http://www.odi.org.uk/sites/odi.org.uk/files/odi-assets/publications-opinion-files/2754.pdf

Ashley, C., Goodwin, H. & Roe, D. (2001). *Pro-poor tourism strategies: Expanding opportunities for the poor* (Pro-poor tourism briefing No. 1). Retrieved March, 23, 2011 from http://r4d.dfid.gov.uk/PDF/Outputs/Mis_SPC/R7557-Brief2.pdf

Ashley, C. & Roe, E. (2002). Making tourism work for the poor: Strategies and challenges in Southern Africa. *Development Southern Africa, 19*(1), 61–82.

Ashley, C. & Mitchell, J. (2005). *Can tourism accelerate pro-poor growth in Africa*? (Opinions 60). London: Overseas Development Institute. Retrieved May, 3, 2013 from http://www.odi.org.uk/opinion/504-can-tourism-accelerate-pro-poor-growth-africa

Ashley, C. & Haysom, G. (2006). From philanthropy to a different way of doing business: Strategies and challenges in integrating pro-poor approaches into tourism business. *Development Southern Africa, 23*(2), 265–280.

Bandyopadhyay, S., Shysamsundar, P., Wang, L., & Humavindu, M. N. (2004). *Do households gain from community-based natural resource management? An evaluation of community conservancies in Namibia* (World Bank Policy Research Working Paper 3337). Washington, DC: World Bank.

Barnes, J. I., MacGregor, J., & Weaver, L. C. (2001). *Economic analysis of community wildlife use initiatives in Namibia* (DEA Research Discussion Paper No. 42). Retrieved from http://www.drfn.info:85/pdf/RDP42.pdf

Blake, A., Arbache, J. S., Sinclair, M. T., & Teles, V. K. (2008). Tourism and poverty relief. *Annals of Tourism Research, 35*(1), 107–126.

Bryceson, D. F. (1999). *Sub-Saharan Africa betwixt and between: Rural livelihood practices and policies* (ASC Working Paper 43). Leiden: De-Agragrianisation and Rural Employment Network, Afrika-Studiecentrum. Retrieved from https://openaccess.leidenuniv.nl/handle/1887/12912

Chirenje, L. I., Chitotombe, J., Gukurume, S., Chazovachii, B., & Chitongo, L. (2013). The impact of tourism leakages on local economies: A case study of Nyanga district, Zimbabwe. *Journal of Human Ecology, 42*(1), 9–16.

Ellis, F. (1998). Household strategies and rural livelihood diversification. *Journal of Development Studies, 35*(1), 1–38.

Ellis, F., Kutengule, M., & Nyasulu, A. (2003). Livelihoods and rural poverty reduction in Malawi. *World Development, 31*(9), 1495–1510.

Ellis, F., & Mdoe, N. (2003). Livelihoods and rural poverty reduction in Tanzania. *World Development, 31*(8), 1367–1384.

Fennell, D. (2008). *Ecotourism* (3rd ed.). London, UK: Routledge.

Freeman, H. A., Ellis, F., & Allison, E. (2004). Livelihoods and rural poverty reduction in Kenya. *Development Policy Review, 22*(2), 147–171.

Gartner, C., & Cukier, J. (2012). Is tourism employment a sufficient mechanism for poverty reduction? A case study from Nkhata Bay, Malawi. *Current Issues in Tourism, 15*(6), 545–562.

Goodwin, H. & Santilli, R. (2009). *Community-based tourism: A success*? International Centre for Responsible Tourism (ICRT) Occassional Paper 11. Leeds Metropolitan University, GtZ. Retrieved March, 24, 2011 from http://icrtourism.org/documents/OP11merged.pdf

Hung, W.-T., Shang, J.-K., & Wang, F.-C. (2013). A multilevel analysis on the determinants of household tourism expenditure. *Current Issues in Tourism, 16*(6), 612–617.

Jones, B. (2001). The evolution of a community-based approach to wildlife management at Kunene, Namibia. In D. Hulme & M. Murphree (Eds.), *African wildlife and livelihoods: The promise and performance of community conservation* (pp. 160–176). Great Britain: James Currey.

Kepe, T., Wynberg, R. & Ellis, W. (2005). Land reform and biodiversity conservation in South Africa: Complementary or in conflict? *International Journal of Biodiversity Science and Management, 1*, 3–16.

Lange, L. (2011). *Exploring the leakage effect in tourism in developing countries* [Diploma thesis submitted at the International University of Applied Sciences Bad Honnef, Bonn, in tourism management]. Retrieved from http://www.giz.de/expertise/downloads/giz2013-en-tourism-exploring-the-leakage-effect.pdf

Lapeyre, R. (2011). The Grootberg lodge partnership in Namibia: Towards poverty alleviation and empowerment for long-term sustainability? *Current Issues in Tourism, 14*(3), 221–234.

Lepp, A. (2007). Resident's attitudes toward tourism in Bigodi village, Uganda. *Tourism Management, 28*, 876–885.

Lepper, C. M., & Goebel, J. S. (2010). Community-based natural resource management, poverty alleviation and livelihood diversification: A case study from northern Botswana. *Development Southern Africa, 27*(5), 725–739.

Mazibuko, S. (2007). Leakages and costs of ecotourism: The case of AmaZizi in the Northern Drakensberg. *Africa Insight, 37*(1), 150–168.

Mbaiwa, J. E. (2003). The socio-economic and environmental impacts of tourism development in the Okavango Delta, north-western Botswana. *Journal of Arid Environments, 54*, 447–467.

Mbaiwa, J. E. (2005). Enclave tourism and its socio-economic impacts in the Okavango Delta, Botswana. *Tourism Management, 26*(2), 157–172.

Mbaiwa, J. E., & Stronza, A. L. (2010). The effects of tourism development on rural livelihoods in the Okavango Delta, Botswana. *Journal of Sustainable Tourism, 18* (5), 635–656.

Meyer, D. (2007). Pro-poor tourism: From leakages to linkages. A conceptual framework for creating linkages between the accommodation sector and 'poor' neighbouring communities. *Current Issues in Tourism, 10*, 558–583.

Mitchell, J., & Ashley, C. (2010). *Tourism and poverty reduction: Pathways to prosperity*. London, UK: Earthscan.

Muchapondwa, E. (2003). *The economics of community-based wildlife conservation in Zimbabwe* (PhD thesis). Kompendiet-Göteborg: Department of Economics, School of Economics and Commercial Law, Göteborg University.

Muchapondwa, E., & Stage, J. (2013). The economic impacts of tourism in Botswana, Namibia and South Africa: Is poverty subsiding? *Natural Resource Forum, 37*: 80–89.

Niehof, A. (2004). The significance of diversification for rural livelihood systems. *Food Policy, 29*, 321–338.

Novelli, M., & Scarth, A. (2007). Tourism in protected areas: Integrating conservation and community development in Liwonde National Park, Malawi. *Tourism and Hospitality Planning and Development, 4*(1), 47–73.

Ogutu, Z. A. (2002). The impact of ecotourism on livelihood and natural resource management in Eselenkei Amboseli Ecosystem, Kenya. *Land Degradation and Development, 13*, 251–256.

Reid, H. (2001). Contractual National Parks and the Makuleke community. *Human Ecology, 29*(2), 135–155.

Rogerson, C. M. (2006). Pro-poor local economic development in South Africa: The role of pro-poor tourism. *Local Environment, 11*(1), 37–60.

Rogerson, C. M. (2012). Tourism-agriculture linkages in rural South Africa: Evidence from the accommodation sector. *Journal of Sustainable Tourism, 20*(3), 477–496.

Sandbrook, C. G. (2010). Putting leakage in its place: The significance of retained tourism revenue in the local context in rural Uganda. *Journal of International Development, 22*, 124–136.

Scheyvens, R. (2002). *Tourism for development: Empowering communities*. Harlow, England: Prentice-Hall.

Shibia, M. G. (2010). Determinants of attitudes and perceptions on resource use and management of Marsabit National Reserve, Kenya. *Journal of Human Ecology, 30*(1), 55–62.

Snyman, S. (2012a). Ecotourism joint ventures between the private sector and communities: An updated analysis of the Torra Conservancy and Damaraland Camp partnership, Namibia. *Tourism Management Perspectives, 4*, 127–135.

Snyman, S. (2012b). The role of ecotourism employment in poverty reduction and community perceptions of conservation and tourism in southern Africa. *Journal of Sustainable Tourism, 20*(3), 395–416.

Snyman, S. (2013). *High-end ecotourism and rural communities in southern Africa: A socio economic analysis* (Unpublished PhD thesis). Cape Town: University of Cape Town.

Snyman, S. (2014). The impact of ecotourism employment on rural household incomes and social welfare in six southern African countries. *Tourism and Hospitality Research, 14* (1–2), 37–52.

Snyman, S. (2017). The role of private sector ecotourism in local socio-economic development in Southern Africa. *Journal of Ecotourism, 16*(3), 247–268.

Spenceley, A. (2003). *Tourism, local livelihoods, and the private sector in South Africa: Case studies on the growing role of the private sector in natural resource management* (Sustainable Livelihoods in Southern Africa Research Paper 8). Brighton: Institute of Development Studies. Retrieved from http://r4d.dfid.gov.uk/PDF/Outputs/Livelihoodsresearch/wRP08.pdf

Spenceley, A. (Ed). (2008). *Responsible tourism: Critical issues for conservation and development*. IUCN, UK: Earthscan.

Spenceley, A., Habyalimana, S., Tusabe, R., & Mariza, D. (2010). Benefits to the poor from gorilla tourism in Rwanda. *Development Southern Africa, 27*(5), 647–662.

Stem, C. J., Lassorie, J. P., Lee, D. R., Deshler, D. D., & Schelhas, J. W. (2003). Community participation in ecotourism benefits: The link to conservation practices and perspectives. *Society and Natural Resources, 16*, 387–413.

Stronza, A. (2007). The economic promise of ecotourism for conservation. *Journal of Ecotourism, 6*(3), 210–230.

Stronza, A., & Pêgas, F. (2008). Ecotourism and conservation: Two cases from Brazil and Peru. *Human Dimensions of Wildlife, 13*, 263–279.

Telfer, D. J., & Sharpley, R. (2008). *Tourism and development in the developing world.* Abingdon, Oxon, UK: Routledge.

Turpie, J., Barnes, J., Arntzen, J., Nherera, B., Lange, G.-M., & Buzwani, B. (2006). Economic value of the Okavango Delta, Botswana, and implications for management. Okavango Development Management Plan. Retrieved from http://www.anchorenvironmental.co.za/Documents/Pdfs/Okavango%20Delta%20Valuation%20FINAL%20REPORT%202006.pdf

United Nations (2015). Resolution adopted by the General Assembly on 22 December 2015 [on the report of the Second Committee (A/70/472)]. 70/193. *International year of sustainable tourism for development, 2017.* Retrieved in November 2017 from http://www.un.org/en/ga/search/view_doc.asp?symbol=A/RES/70/193&referer=/english/&Lang=E

Walpole, M. J., & Goodwin, H. J. (2000). Local economics of dragon tourism in Indonesia. *Annals of Tourism Research, 27*(3), 559–576.

Walpole, M. J. & Goodwin, H. J. (2001). Local attitudes towards conservation and tourism around Komodo National Park, Indonesia. *Environmental Conservation, 28*(2), 160–166.

WCED. (1987). *Our common future. Report of the World Commission on Environment and Development*, G. H. Brundtland (Ed.). Oxford: Oxford University Press. Retrieved November 2017 from http://www.un-documents.net/our-common-future.pdf

Zhao, W. & Brent Ritchie, J. R. (2007). Tourism and poverty alleviation: An integrative research framework. *Current Issues in Tourism, 10*(2 & 3), 119–143.

Tourism development and the empowerment of local communities: The case of Mitzpe Ramon, a peripheral town in the Israeli Negev Desert

Joshua Schmidt and Natan Uriely

ABSTRACT
This study examines the processes of tourism development in the Israeli desert town of Mitzpe Ramon from the perspective of community-based tourism (CBT), a position that emphasizes the empowerment of local communities as a criterion of sustainability. The methodological approach of the study adopts the elements of ethnohistory, in which historical documents are cross-referenced with ethnographic data to provide a portrayal across time that includes the perspectives of both locals (emic) and researchers (ethic). A review of the past six decades of the town's existence reveals that while some of the local communities in Mitzpe Ramon tend to engage in and benefit from tourism development, other local communities are disinterested and/or excluded from this process. This observation complies with the perspective of CBT that local communities are generally derived from a mixture of subgroups with distinct attitudes, interests and social standing. Furthermore, the presented data suggest that while some aspects of empowerment were reinforced due to tourism development, other aspects deteriorated in result of this dynamic. Thus, the study supports the notion of empowerment as a varied and relative construct.

Introduction

The community-based tourism (CBT) perspective suggests that the empowerment of residents is central to the notion of sustainable tourism development (Cole, 2006; Saarinen, 2006; Scheyvens, 1999). In line with this premise, the current study evaluates the aspects of community empowerment associated with tourism development processes occurring in Mitzpe Ramon, an outlying townlet in the highlands of the central Negev desert. The historical review presented in the study is divided across six decades during which the town evolved from a neglected backwater to a tourism destination. This development took place in three main phases: (a) pre-tourism development (1950–1980); (b) development of desert-based ecotourism (1990–2010) and (c) current development of commercial mass tourism (since 2010).

The evolution of the town is depicted below per the impact on and the participation of four seminal local communities: long-standing working-class residents, liberal–secular middle-class newcomers, national–religious middle-class newcomers and Negev Highland Bedouin tribespeople who reside in nearby rural settlements. To assess the degrees of involvement each of these communities have with tourism development and the resulting benefits/costs, the study applies a conceptual model that distinguishes between economic, social, political and psychological aspects of empowerment (Scheyvens, 1999). The empirical research presented in this paper adopts the elements of ethnohistory (Harkin, 2010) in which historical documents are cross-referenced with current ethnographic testimony to provide a portrayal that includes the perspectives of both the resident locals (emic) and the researchers (ethic).

Based on the empirical study case of Mitzpe Ramon, the paper examines seminal arguments of the CBT perspective that view the perception of local communities as mixtures of diverse subgroups with different attitudes, interests and access to power (Clausen & Gyimothy, 2016; Iorio & Corsale, 2014) as well as the notion of empowerment as a relative and multidimensional construct (Scheyvens, 1999). In addition, this paper provides the CBT literature with a suitable methodological approach that relates to the conceptual premise of CBT that claims sustainability concerns the everyday life of residents and that its definition may change across time and phases of development (Saarinen, 2006).

Empowerment and community-based tourism

The current analysis considers tourism development in peripheral areas through the empowerment of local communities. In tourism research, the importance of empowerment is addressed within the conceptual framework of CBT, defined as a tradition in sustainable tourism, where a) the host and the benefits they may gain from involvement in tourism are a central point within the process of development and b) the limits of growth are decided through negotiation and participation (Saarinen, 2006, p. 1129). This approach adds sociocultural and political dimensions to the economic and environmental criteria of sustainable tourism development. As such, the sustainability of tourism development is evaluated per community engagement and support, as well as responsiveness to the needs of the local community. The community is addressed as a mélange of actors with varying attitudes, interests and societal positioning with regard to the scale and type of tourism development (Clausen & Gyimothy, 2016; Iorio & Corsale, 2014; Lew, 1989). The notion of sustainability is considered a fluid social construct negotiated by different actors in the specific time–space continuum of the tourism development domain (Saarinen, 2006).

The concept of empowerment is explained by Cole (2006) as the capacity of individuals or groups within a community to determine their own affairs and to exert control over factors that affect their lives. Specifically, Cole suggests that empowerment is the "top end" of a participation ladder that can span from minimum input and impact at the bottom, to ultimate control at the top. Relating to the same conceptual construct, Scheyvens (1999) examines the tourism development processes via a multidimensional model that distinguishes between economic, social, political and psychological aspects of empowerment.

The profound economic consequences of tourism development are widely discussed in the literature. Scheyvens (1999), for example, suggests that economic empowerment is evident when earnings are shared between several households in the community, whereas disempowerment is apparent when most locals fail to share these economic benefits because they lack capital and/or appropriate skills. Scheyvens similarly posits that political empowerment is evaluated by wide accessibility and inclusion of various groups in decision making, whereas disempowerment is notable when most of the community members feel they have little or no input over the way tourism ventures operate and development unfolds in the region. Hence, political empowerment implies an asymmetrical relationship between the powerful and the powerless, the dominant and the dependent (Sofield, 2003).

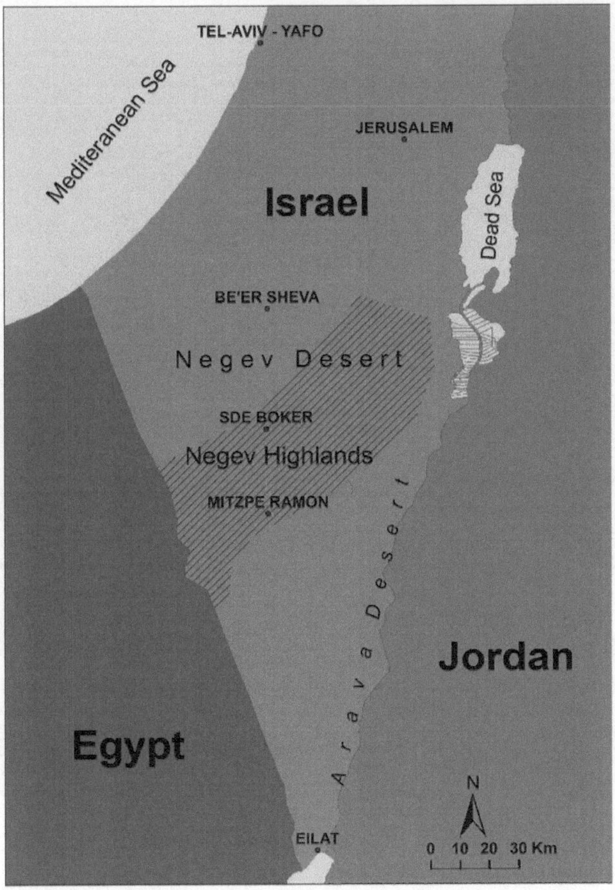

Figure 1. Map of Southern Israel and Mitzpe Ramon environ. Source: the authors.

Social empowerment comprises improvements to community equilibrium and cohesion, whereas disempowerment results in conflict, resentment and jealousy between different local communities. The social dimension of empowerment has been addressed in the context of pro-poor tourism (Ashley, Roe, & Goodwin, 2001), nature-based tourism (Hall & Boyd, 2005) and in relation to specific destinations such as Ngadha, Indonesia (Cole, 2003) and Kiltimagh, Ireland (McGettigan, Burns, & Candon, 2006). Note, however, that the notion of social empowerment without conflict has recently been challenged. Based on an ethnographic examination of the impact of tourism on peripheral rural communities in China, Yang, Ryan and Zhang (2013) employ Coser's theory of social conflict to argue that social tensions and political conflict can be both positive and even necessary for social relationships and a requisite for social change.

Psychological empowerment boosts confidence and self-esteem of many community members who value the external regard and recognition of the uniqueness of their culture and natural resources. Meanwhile, disempowerment is identified when most of the local inhabitants are frustrated, disinterested or disillusioned with local tourism-oriented initiatives. Instances of tourism development initiatives that lead to community psychological empowerment were documented in Kiltimagh, Ireland (McGettigan et al., 2006), and Tanzania (Adler, 1999). The CBT approach provides a fitting theoretical framework for the current case study that focuses on empowerment dynamics among diverse subcommunities in a peripheral town. Accordingly, the Scheyvens' 1999 model of empowerment is the analytical prism through which we examine and evaluate the sustainability of tourism development in Mitzpe Ramon.

Photo 1. Outdoor tourism, Ramon Crater Source: the authors

Mitzpe Ramon and the Negev Highlands

Mitzpe Ramon (population 5,500) is situated in the isolated highlands of the central Negev desert in southern Israel (See Figure 1). The town overlooks the *Makhtesh Ramon* (Ramon crater), a world renown erosion cirque (8 × 42 km) composed of anticlinal ridges that expose exceptional geological formations that are hundreds of millions of years old and normally hidden beneath layers of rubble and vegetation. Located some 90 kilometers south of Beer-Sheva and 150 kilometers north of Eilat, development in isolated Mitzpe Ramon has invariably been tied to the adjacent crater. In the decades following the establishment of the Mitzpe Ramon (1950–1980s), the geological riches in the makhtesh were mined and sold for use in the building industry. Yet, since the 1990s, the Ramon crater and the surrounding Negev Highland region transformed into a site of ecotourism (Mazor, 1992; Reichel & Uriely, 2003; Reichel, Uriely, & Shani, 2008; Schmidt, 2015; Uriely, Reichel, & Shani, 2007). Consequently, during this period local mining operations decreased, while ecotourism activities expanded into a comprehensive tourism industry.

Currently, tourism attractions in the Negev Highlands include heritage sites, national parks and activities associated with outdoor adventure, spa, wellness and alternative health (Photo 1). The central clientele for these services are domestic tourists and, more recently, visitors from abroad as well as day excursionists to Mitzpe Ramon and the makhtesh countryside. Although a few of the tourism operators provide services to organized package tour groups, they mainly focus on independent travelers who favor self-arranged transportation and accommodation (Hyde & Lawson, 2003).

Recent research on rock art tourism development in the Negev Highlands assembled the limited available tourism statistics, finding that in 2013 the area received nearly 1,300,000 combined visitors to heritage and nature-related sites (IMPART, 2016). Incoming tourists comprised 35%, while 65% were from the domestic tourism circuit. Hotel rooms comprised 56% (6 hotels/363 rooms) of all accommodation for the Negev Highlands. The remaining rooms were divided between youth hostels and field schools (23%, 147 rooms in 3 sites) and rural tourism accommodation (21%, 135 rooms). Camping sites contained another 3,400 beds that are mainly used for domestic tourism and incoming young-adult independent tourism. In 2013, the average annual rate of room occupancy in tourist hotels in Israel was 66% and 70% in the Negev Highlands. Future plans promoted by local and national governmental agencies include large-scale investment for renovation of existing tourism infrastructure, development of new attractions and construction several luxury hotels along the edge of the crater.

Methods

The methodological approach of the study adopts the elements of ethnohistory (Harkin, 2010) in which historical documents are cross-referenced with ethnographic data to provide a portrayal that includes both the local (emic) and researcher (ethic) perspectives. The article depicts the historical development of the town from its inception in the 1950s to the present day based on archival data, especially newspapers and government surveys. The official narrative was verified with extensive ethnographic research conducted over the past decade among the aforementioned four local communities.

Ethnographic fieldwork took place in and around Mitzpe Ramon throughout the past decade when one of the authors resided in town. During this period, the researchers grew familiar with assorted members of the four core communities, whose combined numbers represent some 80% of the local population. Fieldwork included participation observation at the ongoing formal and informal gatherings and daily life events of each of the communities. Fieldwork also occurred digitally, with the authors engaging in resident community web-based social networks. During the initial stages of fieldwork, the authors identified and categorized the central features of each community. They then drew up a list of questions pertaining to local development and proceeded to conduct fifteen uniform interviews per community with a mixture of male and female informants including principal community figures and general, private residents who were mentioned in conversation and/or had an active online presence. In addition, the authors held numerous ad hoc discussions and had regular daily interactions with informants of all ages, economic conditions and sociocultural backgrounds whose pronounced worldviews were assessed vis-à-vis the opinions expressed in the more official interviews and online posts.

Additionally, the researchers conducted semiformal interviews with nearly fifty (out of an estimated 175) private sector Negev Highland tourism operators working in the fields of hospitality, caregiving, guiding, transportation, marketing and management services. The interviews were conducted as discussions in which the researchers probed informants, often fellow town residents, on certain topics, issues or areas of interest associated with the study. Interviews took place either onsite, within a tourism work context, or at popular residential venues. Interviewees were generally between the ages of 25 and 55. Roughly half were married or otherwise settled with their families in Mitzpe and about one-third were female. While their years of experience in the field differed, the majority of the informants earned their primary living from tourism-related occupations.

The researchers also conducted twenty formal Key Informant Interviews (KII) with representatives from the public sector. Informants included former and current district and central government role holders from the slew of offices, branches and departments linked to regional development. Interviews took place over several years of study and were often arranged via a "snowball" methodology in which one interviewee recommends additional informants whose names oftentimes, particularly following an online search, led to a broadened understanding of the research field. Interviewees were asked a series of pre-established questions that were revised ongoing per insights from the preceding fieldwork. To encourage cooperation, the researchers refrained from digitally recording their KII dialogues. Instead, they took notes of key points and statements that they later fleshed out from memory. To retain authenticity, interviewees were chosen per professional relevancy to the different segments of the research and thus did not necessarily know one another, hold top-level positions or harbor a comprehensive acquaintance of development in the Negev, but rather related to specific issues per their own specialized understanding.

The researchers performed a content analysis of the assembled data, identifying key issues and central themes located in the verbal (spoken) and written (online) language of their informants. Viewing language as an instance of human behavior that contains both communicative and cultural functions (Tobin, 2009), the researchers applied a sign-oriented anthro-linguistic approach to identify both linguistic *and* extralinguistic messages in the data, thereby establishing an empirical link between the theoretical knowledge and observable behavior. The ethnographically informed linguistic analysis of words, phrases and expressions connected to the diverse communities helped the researchers determine the underlining emic-based messages contained within the language of the different communities.

First-hand familiarity with the field permitted the researchers to navigate the gaps between the objective linguistic data and the subjective attitudes and personal viewpoints expressed by the respective informants. What emerged was a discourse riddled with dichotomies, contrastive wording and expressions that seemed systematically reflective of the assorted degrees of empowerment experienced by the distinct communities. The analysis of the data assembled from the KII's and associated ethnographic fieldwork provided a clearer understanding of the intersecting economic, sociopolitical and institutional forces – and the particularized local reactions to them – that steer the course of tourism development in the region. The cross-referencing of sources – official and informal, public and private, historical and current – and multiple modes of analysis informed the examination by enhancing the internal validity and reliability of the research findings.

Phase one: Pre-tourism development (1950–1980)

Following the founding of the State of Israel in 1948, the town that is today known as Mitzpe Ramon was initially established as an army outpost. In the early 1950s, the site also served as a base for teams of workers who were installing national infrastructure in the region. In 1956, some twenty families arrived onsite, seeking to initiate an "urban-agricultural cooperative" (Elishav, n.d.) that they named Mitzpe Ramon ("Ramon Lookout"). As its economic base, the cooperative relied on the mining of gypsum and other raw mineral deposits in the crater basin that were sold to the national building industry. Yet, unforeseen logistical challenges and bureaucratic impediments led to financial loses and subsequent interpersonal tensions. While the cooperative lasted less than two years, the outlook its members held toward the Ramon crater – as a source of material revenue – set a precedent and remained the mindset of subsequent residents in the ensuing years.

In 1958, the National Council for Immigration and Absorption assumed administrative control of Mitzpe Ramon. Incorporating the town into its web of government-supported development towns, the ensuing first development phase was underwritten by the central government. "Mitzpe", as it became known by resident locals, became part of a network of thirty-two outlying settlements setup in the early period of the state to accommodate the hundreds of thousands of Jewish immigrants, who in the early 1960s were pouring into Israel, particularly from North Africa and Middle Eastern countries. Collectively known as the *Eidot Hamizrach* ("Tribes of the East") or colloquially as *Mizrachim* ("Easterners", *Mizrachi* in singular), most of the immigrants assigned to live in Mitzpe Ramon by the government arrived from assorted localities in Morocco, either directly off the boat or from transit camps in the center of the country.

The classic perception of "frontiering" as the heroic hard-won settlement of marginal lands in advance of an organized society (Turner, 2008) fails to describe the experience of the early Mizrachi residents in Mitzpe Ramon. In the years following the founding of the state, Euro-Anglo-oriented Ashkenazi attitudes and ideals dominated Israeli behavior and culture. Responding to their ethnic otherness, the mainstream Ashkenazi establishment relegated the Mizrachim to the institutional, intellectual, social and geophysical peripheries of the Jewish-Israeli society (Zameret, Halamish, & Meir-Glitzenstein, 2009). The Mizrachim, especially residents of the so-called development towns, were stigmatized. Perceived as "Jewish Arabs" who were poor, backward and detached from modernity, their values and perspectives were viewed as secondary to those held by the native-born Israelis living in the more affluent center of the country (Gutwein, 2004). Possessing neither the necessary political clout to reverse their marginalization, nor the private capital to generate investment and self-betterment, the residents of Mitzpe Ramon lingered on the margins of the Israeli periphery. Disenchanted and embittered by their marginal presence within the country and unattached to the socialist ideals and Zionist pioneering ethos of the earlier Euro-Anglo settlers, many of these veteran immigrants remained bitter about what they perceived to be discriminatory circumstances that brought them to live in the remote Negev Highlands. Accordingly, they were defined by Weingrod (1966) as "reluctant pioneers" (Photo 2).

To be sure, the national government recognized the need for progress and enacted certain development strategies to improve the conditions of the Negev residents. Yet, as one veteran immigrant resident acerbically described it, "funny, the officials in Jerusalem were planning and developing Mitzpe to suit *their own* needs" (Y. Ben-Gigi, personal communication, November 22,

Photo 2. Formerly from Morocco, longstanding Mitzpe resident in front of his homemade menagerie. Source: the authors

2013). With government initiatives subject to the alternating agendas of competing political parties, they tended to lack foresight, a comprehensive vision or consistent management. Moreover, regional development operations were void of local input and failed to include substantial resident collaboration in their design or implementation (Rosenhek, 2011). Trivializing resident sensibilities, the government-funded initiatives were largely managed from afar and therefore poorly executed and without proper mechanisms to ensure administrative or logistic follow-up on both national and district levels (Plieninger, Dijks, Oteros-Rosaz, & Bieling, 2013).

Settling into their sidelined status, the working-class *Mizrachim* assumed an ambiguous cultural affinity with the rest of the country. Responding to the ineffectual government development programs, the town inhabitants preserved societal mores from their former countries of residence as a way of offsetting their existential exclusionary present. In result, during this period, local culture was infused with the opinions, beliefs and mannerisms of the local Jewish-Moroccan immigrant residents (Hoffmitz, 1978). This era was perceived by many of the long-standing locals we interviewed as a period of endearing communality when, in result of their isolation, the entire resident population "personally knew one another and no one bothered to lock their doors" (M. Deri, personal communication, November 22, 2013).

As a government-supported development town, during this initial phase many locals were employed as day laborers in the municipality or for mining companies or found jobs in national construction projects in the region. Others worked at the nearby military bases and prisons or in government-contracted factories. A few operated the basic provisional shops about town and only a handful of local citizens were gainfully self-employed or otherwise commercially independent. Moreover, a proportionally large segment of the resident population relied on government unemployment benefits and other means of social welfare assistance to support itself.

In the 1970–1980s, stagnant growth and insufficient economic opportunities triggered a process of reverse emigration from Mitzpe Ramon and the resultant process of negative selection (Ben-Zadok, 1993). Residents with marketable skills and/or social connections elsewhere in Israel left in search of improved employment conditions and broader education opportunities for their families. In consequence, the town lost much of its social and intellectual capital. The situation further deteriorated in the mid-1970s when the highway linking the north of the country with Eilat was diverted through the Arava desert, depriving Mitzpe Ramon of the financial benefits it previously received as a stop-off point for vehicles travelling south. "It was as if the country had forgotten about us", recollected a long-time resident about that era (I. Edri, personal communication, July 28, 2013).

Since mining in the Ramon crater persisted and the general Mitzpe locality was branded with a lowly socioeconomic standing, this phase saw meager tourism activity. To the degree that inns, hostels or other tourism enterprises existed, they catered mainly to school and national youth movement trips or rugged in-country visitors, or else offered temporary residence to laborers working in the area. Long years of institutional disregard coupled with general indifference toward the plight of the distant inhabitants ultimately rendered Mitzpe Ramon – whose remote location was and continues to be its central attribute – a neglected government-dependent settlement that, in fact, never fully developed.

Despite two major state-directed demographic interventions – namely, the relocation of army bases from the Sinai Peninsula into the Ramon region subsequent to the 1979 Camp David Peace Accords between Israel and Egypt and, in the early 1990s, the arrival of nearly a thousand Jewish immigrants directly from countries in the former Soviet Union – Mitzpe Ramon remain in its stalled condition until the turn of the century when converging newcomer groups began supplanting the local economy and residential culture with their distinctive worldviews and agendas. It should be mentioned that during this initial period, additional subgroups resided in Mitzpe concurrent to the main Moroccan community, e.g. military service personnel, scientific researchers, artists and retirees. However, their comparably minimal size, transient nature and attachment to external sources of income and culture meant that, unlike the veteran Mizrachi community, these groups sidestepped the local narrative of systematic marginality and public indifference, preferring a more affirmative view of their peripheral conditions (Rapport, 1998).

Phase two: Desert-based ecotourism development (1990–2010)

The rise of the Israeli environmental movement in the 1980s and 1990s (Tal, 2002), coupled with the recognition of the economic viability of ecotourism ventures in the desert, altered the perceived value of the Ramon crater. Environmentally oriented newcomers transformed the Ramon crater narrative by promoting a local tourism industry centered on its continued preservation rather than the extraction and sale of its raw materials. With the makhtesh environment proving valuable for its sightseeing rather than industrial cache, mining was discontinued as "crater tourism" became central to the Mitzpe economy.

The tourism ventures to emerge in this phase were compliant with the defining principles of ecotourism, wherein travel to natural areas aims both to be educational, environmentally sustainable and generate benefits for locals (Jones, 2005; Weaver, 2005). Phase two in the evolution of Mitzpe Ramon is depicted below with respect to three local communities: (a) liberal–secular middle-class newcomers, mostly of Ashkenazi (Anglo-European) ethnic origin; (b) national–religious middle-class newcomers, who are also predominantly Ashkenazi; and (c) Arab Negev Highland Bedouin, who live in unrecognized villages throughout the adjacent region.

a) Liberal–secular middle-class newcomers

The local liberal–secular middle-class newcomer community began taking shape in the late 1980s and gained form throughout the 1990s and early 2000s. This group is primarily composed of families and singles who moved to Mitzpe seeking respite from the accumulative tensions stemming from the Arab–Israeli conflict and the socioeconomic pressures connected with contemporary Israeli urbanity. Having chosen to live in the periphery, they bettered their sociomaterial conditions by taking advantage of certain factors found in Mitzpe at that time, such as a small population, low cost of living, reduced income tax, minimal competition, the absence of commercialism and slow pace of life. Being culturally and financially self-sufficient, members of this community generally professed to being oblivious to the town history and the condition of its other, more permanent resident communities with whom they have narrow social and cultural interactions or personal connections.

The liberal–secular middle-class newcomers maintain a contextual appreciation of Mitzpe Ramon that vastly differs from how many of the long-term, assigned working-class residents envision the town. Unlike the veteran residents, the newcomers we spoke with claimed to have made a personal choice to relocate to the desert in pursuit of individual, rather than ideological or communal, goals. Celebrating the "rural idyll" of their desert isolation (Mingay, 1989), they bask in the primal natural landscapes and romantically interpret their presence in the primordial hinterland. Corresponding with what Shields (1991) terms "near-sacred liminal zones of otherness", the newcomers claim a fascination with the sweeping, uncluttered landscape and draw spiritual inspiration from the pristine natural environment found in the crater. However, many are not committed to living in Mitzpe for the long term. Instead, community members often describe Mitzpe as a "stop on the path" and perceive the town similarly to what Nielsen and Nielsen (2016) refer to as a "free space" in which sanctified desert scenery and time–space metaphysical marginality combine to become available, if transitory, sources of wellness.

The influx of the comparatively prosperous northerners into the barren Negev is related to the fragmentation of traditional Zionism and the resultant post-Zionist cultural turn that evolved in Israel in the wake of the dispiriting 1973 Yom Kippur War (Klein Halevi, 2013). As the country underwent a foundational ideological shift, its collectivist underpinnings morphed into material driven individualism. Communal service and personal sacrifice for the nation, previously the hallmarks of the Israeli cultural imperative, were replaced with the quest for individual gain and globally informed personal self-betterment (Ben-Porat, Levi, Mizrahi, Naor, & Tzfadia, 2008; Ram, 2000; Shapira, 2004).

One effect of this transformation, especially among the younger generations, was that it became a standard, even expected, rite of passage for Israeli youth recently discharged from obligatory military service to venture to far-flung localities overseas to "find themselves" amidst the quieting scenery and reduced living costs to be had at Israeli tourism enclaves (Cohen, 2008; Noy & Cohen, 2005). As with other rites of passage (Turner, 1982), during their time away from mainstream society, Israeli backpackers assume unique roles and undergo transformative experiences that alter their status upon reemergence in daily society. The arrival of the secular middle-class newcomers to Mitzpe Ramon is part of this same self-serving disengagement–reengagement process. That is, either separately or in conjunction with their overseas trips, many young Israelis choose to withdraw from the sociopolitical and economic intensity of their lives by temporarily residing in the Mitzpe enclave where they focus on identity experimentation, communing with the natural surroundings, themselves and each other.

The individualist slant of the liberal–secular newcomer community is readily detected in the language of its members. For instance, a hand-painted sign, written in English and located at a lookout point on the edge of the crater, reads "Wherever you look, see yourself" (Photo 3). In another example, the self-absorbed "I am from Mitzpe" was the catchphrase of the newcomer political party in the 2013 municipal elections. Instead of applying the plural/collective "we", as in "we are all from Mitzpe", the language invokes the singular/individual "I", implicitly conveying the message: "my residency in Mitzpe is a privileged, individual choice". Validating this outlook, the recently elected newcomer mayor launched a tourism marketing campaign with the slogan "Mitzpe Ramon: A Source of Inspiration".

Photo 3. "Wherever you look, see yourself", hand-painted sign at Ramon Crater lookout point. Source: the authors

During the second development phase, the newcomer tourism service operators fused the crater and town into a single product. Localized tourism ventures were largely based on "do-it-yourself" labor (Reichel & Uriely, 2003). Utilizing available spaces around town, enterprising newcomers erected an array of funky shops, restaurants, bars and coffee houses, arts and crafts, music and culture studios, frequently in refurbished industrial hangars. Several built farms outside of town and invested in artisanal wineries, creameries, petting ranches and a "silent" bow and arrow range. Assuming a grassroots character, local tourism services included "green" B&Bs, off-the-grid camping huts, home-cooked catering, alternative medicine and desert herbal remedy treatments, backcountry tour guiding, star-gazing and jeep tours, rappelling, horseback, motocross and bike riding.

The homespun character of the Mitzpe Ramon tourism experience gained a positive reputation for its quaint authenticity and affected a fashionable buzz in the national and international tourism presses with the town being touted as the "capital of Israeli desert tourism" (Rivelli, 2016; Toll, 2016). While the newcomers capitalized on the packaging of their local desert habitat, for the most part, the more permanent residents had scant involvement in the development activities and gained little economic, social or political empowerment from these operations. As they grew in size and influence, the liberal–secular newcomer perception of Mitzpe – immersed in eco-spirituality and attuned to cosmopolitan middle-class material needs and emotional sensibilities – eclipsed the former narrative of disregard, mismanagement and neglect held by many long-standing working-class residents.

Resourceful, confident and well-informed, the secular newcomer community proved both socially and financially adept in developing Mitzpe Ramon tourism, in part because its members could contend with the complex structural obstacles that have historically hindered the fundamental advancement of the town. As to be expected, newcomer tourism initiatives generally refrained from engaging with the sociocultural history of the town and its older resident populations. Instead, Mitzpe tourism showcased the surrounding natural environment and the hip newcomer enclave of workshops, eateries, galleries, bars and lodgings. In consequence, akin to the newcomers themselves, most tourists to Mitzpe Ramon are oblivious of the town's earlier narrative of displacement and stalled growth.

b) National-religious middle-class newcomers

The national-religious (also known as religious Zionist) community in Mitzpe Ramon combines traditionalized Orthodox practice with messianic driven nationalism. This group is affiliated with a nationwide Torah-inspired education-oriented settler network known as "*Garinim Toranim*" (lit. Torah kernels; *Gar'in Torani* in singular). The groups in the network act as a countermeasure in areas of Israel where Torah observance is culturally lax or politically threatened (Reichner, 2013; Taub, 2010). Ideologically inspired, garinim toranim concentrate their strength-in numbers to fortify and expand national–religious doctrines, often within unsympathetic civic environments.

Although both the national–religious and liberal–secular newcomer communities arrived on their own accord, their motivations for moving to Mitzpe were quite different. The secular newcomers sought personal betterment and individual fulfillment. The national–religious gar'in aspires to actuate divinely mandated and collectively executed nation building. Thus, the gar'in first came to Mitzpe in February of 2000 and immediately established a Torah seminary. While the nearly 300-person student body at the seminary fluctuates, the teachers and their families are the stable foundation of this community. With over fifty affiliated families, the gar'in members make a point of purchasing homes and establishing themselves as permanent residents in the neighborhoods adjacent to the seminary. "We came here to make – with the help of God – a lasting difference" is a sentiment regularly articulated by gar'in informants (S. Katz, personal communication, July 28, 2014). At the same time, the gar'in was in part attracted to Mitzpe

Ramon because its remote setting imbues a monastic quality that enables members, particularly the seminary students, to remain apart from what they consider debased mainstream contemporary culture.

Organized, focused and highly motivated, the gar'in is quasi-self-sufficient having established an internal network of community platforms including educational, economic, political, cultural, housing, sociocommercial and media-information (Photo 4). Many community members are employed in occupations connected with gar'in undertakings and predominantly work in education, social services and caregiving. Engaged with their Orthodox life-worlds – daily Jewish observance and an endless roster of sociofamilial obligations and activities – gar'in members rarely partake in non-communal or non-Torah oriented events, nor do they maintain significant personal relationships with people outside the national–religious cultural sphere. Yet, in line with the objective of actively disseminating Torah – and in contrast to the secular newcomers – this community makes a conscientious effort to be in touch with the less privileged local communities and sponsors several outreach programs. As indicated below, some of these programs comply with the emphasis given to socially oriented initiatives of ecotourism (Jones, 2005), including pro-poor tourism that aspires to alleviate poverty and other social ailments afflicting host communities (Bowden, 2005; Torres & Momsen, 2004).

Arriving in the middle of the second development phase, the gar'in primarily concentrated on establishing its own institutions and did not play a major role in the emerging tourism development. Moreover, while the tourism initiatives of the liberal-secular community are underlined by their commercial nature, gar'in informants told us that they view personal profit as secondary to the Torah-construed value inherent in the act of their settlement project. Thus, to the extent that gar'in members pursue for-profit ventures, these activities are circumscribed within a precise

Photo 4. National-religious seminary students circle dance during the inauguration of their education complex, Mitzpe Ramon. Source: the authors

financial and contextual scope as, in effect, they are considered extensions of established religious directives. For instance, the gar'in operates an apartment-hotel guesthouse (fulfilling the edict of "hospitality to strangers") and a low-cost car rental service that runs on minimal profit margins (fulfilling the edict of "mutual aid").

Another example is *Kerem Ramon* ("Ramon Vineyards"), a vineyard planted on the outskirts of town that is owned and operated by gar'in members (fulfilling the edict of "settling the land"). Kerem Ramon was founded by fourteen families from the gar'in within the context of the *Sfat Midbar* ("Desert's Edge") framework, a government-endorsed development program centered on agritourism entrepreneurship. While several secular entrepreneurs set up private, for-profit farms, Kerem Ramon is perceived by its affiliates as a platform for fulfilling the precept of being charitable and concerned for one's fellow beings. This ideal is confirmed on the Kerem Web site whose "About us" statement claims that the gar'in will "donate fifty percent of the net profits from the vineyard to be used by the local community for education and welfare programs" (http://keremramon.org).

The language used by our gar'in informants during interviews and in social media platforms similarly underlines the ideological reluctance this group harbors toward overt material gain and its penchant for lending support and assistance to the local community. Gar'in posts are written in a mater-of-fact style, contain numerous acronyms for religious terms denoting togetherness and combined effort and avoid abusive language, immodest subject matter and opinionated gossip. Yet, despite their lack of direct alignment with the commercial aspects of the Mitzpe tourism industry, during this phase, the singularly determined gar'in members accumulated significant influence over local politics and advertently contributed toward advancing the third phase of local tourism development.

c) Negev Highland Bedouin

The Bedouin tribespeople living in the Negev Highlands are Arab-Israelis who reside in the Mitzpe Ramon vicinity and constitute an additional community addressed in the current study (Photo 5). The presence of the Bedouin in the regional Arabian deserts goes back at least two millennia. As they historically resisted ordinance, over the past five-hundred years, the sequence of Ottoman, British and Israeli administrators considered this group to be something of an administrative black hole (Marx, 2012; Meir, 1997). The succession of governments variously mishandled the Bedouin – who were settled and resettled, uprooted, transplanted, relocated and finally urbanized – to suit varying, and at times contradictory, short-sighted national interests and needs.

The sometimes abrupt policy swings, particularly those enacted by the Israeli government over the past few decades, have caused large portions of this community to be wary of present-day government endeavors to incorporate them within mainstream socioeconomic structures and systems. Current government figures suggest that nearly forty-five percent of the Bedouin (circa 110,000 people) are living in unplanned villages that are considered "illegal" by the Israeli government and therefore "unrecognized" in terms of entitlement to the basic government-supported infrastructure and services (Kedar, Amara, & Yiftachel, 2018). This is the situation with the Negev Highland Bedouin, with roughly a dozen extended families inhabiting five unrecognized villages along Route 40, the highway that runs between Mitzpe and Sde Boker, the nearest settlement some thirty kilometers to the north (Figure 1).

The Negev Highland Bedouin, all of whom are full-fledged Israeli citizens, are unique in that their remote-living circumstances have enabled them to cling to certain aspects of their tribal past, particularly herding and lodging in extended family encampments. Existing in disparate states of socioeconomic squalor, the estimated one-thousand tribespeople reside just beyond the borders of the Mitzpe Ramon municipality on what they claim to be their ancestral lands,

Photo 5. Negev Highland Bedouin tribeswoman during wheat harvest. Source: the authors

today located within the contiguous Ramat Negev municipality. This community professes a primal connection to the Ramon environment. Several Bedouin elders can even recall being born in the Ramon crater basin where informants recounted in interviews, prior to the establishment of the state of Israel their families camped for the winter, a practice the national authorities halted in the 1950s. As a cultural ideal, an intimate connection with the desert continues to be an essential component of Negev Highland Bedouin identity. This trait sets them apart from more urbanized Bedouin communities in central regions of the Negev who have ceded certain components of their traditional lifestyle, deeming them outdated or incompatible with their contemporary environs.

Negev Highland Bedouin of all ages frequent Mitzpe Ramon to attend to their basic commercial, medical, administrative, employment and educational needs. Community members maintain an affable working relationship with the other town residents. Adult males are frequently employed as menial laborers for the municipality, the National Parks Authorities or at service stations and construction sites in the region. Belonging to a disenfranchised Arab minority, few, if any, volunteer for the army as in the past. Despite their close physical proximity, the Bedouin and the other Mitzpe Ramon communities exist in disparate life-worlds and seldom interact with one another on a sociocultural or political-economic level.

Recently, the Negev Highland Tourism board has been instrumental in lobbying to settle the Bedouin communities through rural tourism initiatives. Coined the *Shvil Ha'Ohalim* ("The Tent Trail"), the plan calls for the Bedouin to host visitors during the day in their currently unrecognized encampments and return home for the night to a single, recognized township (public hearing, Sde Boker, March, 2014). Through the production of neotraditional Bedouin tourism, the Shvil Ha'Ohalim initiative provides the Bedouin with a suitable platform for gainful employment while also enabling them to maintain an approximation of their traditional habitat. Moreover, the plan satisfies government concerns regarding the long-term status of current unrecognized

Bedouin villages by making a clear distinction between official daytime sites for Bedouin hospitality and permitted places of fulltime residence.

In expectation of the official approval of this model, local tourism operators are working with Bedouin partners to host visitors in their villages and the surrounding region. These activities are especially favorable for the newcomer entrepreneurs who promote a romantic narrative of Bedouin history and culture and manage the supply and demand channels and the division of the earnings. Yet, to some extent the Bedouin are likewise benefitting from this enterprise, having recognized the potential for financial gain by presenting a fabled version of their heritage within a tourism context. Thus, despite their neocolonial underpinnings, the Bedouin–newcomer partnerships have generated the basis for the integration of Bedouin tourism within future regional development plans.

Although they would prefer to remain in their current locations, several of our Bedouin informants acknowledged that their communities have "moved 500 years in [the last] 50 years", so that now they "use tourism to preserve what's left of our roots" (S. Sadan, personal communication, March, 2014). At the same time, urged by their radical tribal confederates in more central parts of the Negev, the local Bedouin community has so far declined to openly support the *Shvil Ha'Ohalim* proposition, as this stance would effectively constitute a waiver of its direct usage and ongoing connection with present land holdings. Meanwhile, the army, a principal representative of the central government in the region and the registered property holder of the lands upon which most of the Negev Highland Bedouin reside, has lately voiced opposition to the Bedouin tourism initiative. At a meeting of the southern district regional planning committee in Beer-Sheva, military officials emphasized the impermanence of the five Negev Highland Bedouin villages declaring that, with one exception, they are "located within active firing ranges and therefore legally off-limits to all civilians" (D. Gur, personal communication, May, 2015). This attitude is demonstrative of the systematic reluctance of the government to lend legitimacy to Bedouin land ownership claims or adequately include them in decision making processes connected with local development. Accordingly, although the Bedouin have lived in the area longer than any of the other resident communities, their reduced political stature, manifested in their lowly socioeconomic conditions and perpetuated by a chronic state of civic impermanence, ultimately excludes them from fully participating in regional tourism development processes.

Phase three: Present and impending commercial tourism development (since 2010)

As they evolved from newcomers to long-standing residents and invested more deeply in their respective ventures, the two middle-class communities also began taking an interest in Mitzpe politics. After a decade of failed attempts to join the city council as separate parties, for the 2013 municipal elections the newcomer communities united to form a single party. A central component on their agenda was the removal from power of the veteran community to make way for transfiguring Mitzpe Ramon into what they referred to as a "better place". Led by a former army colonel turned secular newcomer, the party pledged to promote local demographic expansion and ecotourism development. The newcomer party won the election by a landslide, clearly indicating that the local tides were changing.

In the months following their victory, the triumphant newcomer party members undertook a fundamental reorganization of the local municipality, purging its former working-class core and initiating their election promise to invigorate town governance. Nearly four years since assuming power, the key education, community development, business and social welfare positions at the local municipality are staffed by secular and/or religious newcomers whose communities have also seen a surge in funding for their community-oriented education and social events.

Encouraged by the invigorated city council, the enterprising newcomers are attracting like-minded affiliates from other parts of the country to join in the local expansion efforts. Transformation in Mitzpe Ramon is evident which is abuzz with activity. The municipality is enthusiastically formalizing the management and development of the town with intentions to expand its population size by nearly forty percent. Underway are numerous building projects, including four new neighborhoods, a shopping mall and collaborative office space for small businesses. Municipal workers have installed a plethora signs, speed bumps, gates, fences, recycling bins, street lights, road signals about town and assigned designating parking areas and traffic squares. At night, the streets are illuminated with blinking solar-powered LED hazard lights, digital speed detectors and the luminous glow from spotlights shining down upon a new skatepark in the center of town. Furthermore, at a town assembly in November 2017, the municipality presented plans for the construction of a mega neighborhood and accompanying public infrastructure along with a second business–industrial area that, if enacted, will double the size of the town and likely create a kind of neo-Mitzpe Ramon.

It is anticipated that the burgeoning scope of economic investment over the next decade will greatly accelerate tourism development. Recently, the Israeli Ministry of Tourism declared its intention to invest 65 million US dollars in the next several years to revamp the tourism infrastructure in Mitzpe Ramon. Concurrently, the Israel Lands Authority announced that it would soon begin marketing land holdings on the outskirts of town for the erection of six high-end luxury hotels. Thus, the apparent actors in the emergent third phase of local tourism development are incoming large-scale commercial developers who are set to wrought profound changes to the town and the surrounding area. The unprecedented scale of fiscal and physical investment and subsequent increase in tourism volumes will certainly have significant repercussions on the town character, as current residents will be exposed to a convergence of part-time hotel service workers and varying *new* newcomer investors and their associated socioeconomic circles.

Presently, the most prominent example of the new style of corporate investment in Mitzpe Ramon is the Beresheet hotel. Part of the Isrotel international hotel chain, Beresheet overlooks the Ramon crater and, at nearly 700 dollars per night for the cheaper rooms, is considered one of the best and most expensive hotels in the country. In the early 1990s, during the second, more modest phase of tourism development, Isrotel successfully converted an unoccupied residential housing project into a three-star hotel that, at the time, mainly employed locals. With the realization of this enterprise, the chain came to understand the potential for upper market tourism in Mitzpe Ramon and began drafting ambitious plans to build a five-star hotel overlooking the crater. Two and a half decades later, the socioeconomic impact the ambitious Beresheet project is having on Mitzpe Ramon is double-edged and therefore serves as an exemplary instance of what might transpire from the subsequent development efforts on this scale.

On one hand, roughly a third of the hotel staff are resident locals who are mainly employed in minimum wage service jobs. The hotel hires most of its managers and trained staff from out-of-town and employs a smattering of foreign migrants who are provisional, and thus, their effect on local demographics is so far undetermined. In addition, citing the need for reliable professionalism, Beresheet tends to bypass local service providers, preferring instead to outsource its in-field operations to practitioners from the center of the country.

While the chain has embarked on a national advertising campaign that presumably aids the overall branding of Mitzpe, ultimately the revenues from this enterprise are garnered by Isrotel executives in the center of the country and overseas. Furthermore, the hotel does little to integrate itself with the local tourism activities. Many residents we spoke with, as well as Negev Highland tour operators, town officials and the Isrotel executives themselves, consider Beresheet to be what they referred to as "an island to itself". To be sure, with the hotel and its services being promoted to the wealthy elite, guests generally stay within the Beresheet complex and commonly see very little of the town.

Since its opening in 2011, the hotel is running at close to full capacity. The researchers found little evidence to suggest that Beresheet is actively preventing its environmental footprint from harming the surrounding ecosphere. Operating a hotel of this scale means inflated vehicle and human traffic, excessive water use and the production of waste – actions that are potentially ecologically harmful but will likely only be assessable after the fact. Indeed, the business of Beresheet is business and with the hotel headquarters located elsewhere, its effect on the immediate physical and/or social environments seems of secondary concern to its owners and managers. Accordingly, as a precursor of local corporate business, the extent of its gainful contribution on the town is questionable, requiring additional inquiry.

As might be expected, many of the more veteran newcomers are reconsidering their support for the course the larger-scale development is assuming. This is particularly true for those individuals who have lived in Mitzpe for a decade or more were the initiators of the localized ecotourism industry and originally endorsed the political change and the newcomer party strategy for socioeconomic development. Fearing the loss of their idyllic lifestyle, many liberal–secular newcomers are skeptical of the ability of the town to accommodate external investment without compromising its unique character and singular natural habitat. "No one believed change would happen so fast… and so extensively", decried one of our newcomer informants, a tourism operator and erstwhile supporter of the newcomer party, remarking on the concerted pace and scope of development and the potential for permanent environmental damage (G. Davni, personal communication, December 19, 2016). This sentiment is echoed by many of the tourism operators we spoke with who critiqued what they perceive as misguided plans to (over)develop the town and, in the process, hallow out the authentic appeal of the local tourism industry.

The expected advancement of the third phase of extensive tourism development is also being critiqued by some of the national-religious community who are concerned that the growth of the Mitzpe tourism industry – whose leading activities transpire on the Sabbath and Jewish holidays when Orthodox Jews refrain from working or conducting business – will inevitably lend the town a stronger secular character. Until now, tourism activities occurring on the Sabbath have been low key and there exists an implied status quo wherein, as per the preference of its initial traditional Mizrachi residents, the businesses in Mitzpe Ramon close for the Sabbath and Jewish holidays. With the arrival of corporate-owned businesses, there is apprehension that their inevitable operation on the Jewish day of rest will spoil the currently maintained commercial-free Sabbath ambience. At the same time, a cross section of both new and veteran secular and religious residents have expressed ambivalence toward further development citing recent increases in garbage, traffic and the prices of local goods. To be sure, with the municipality laboring to balance the town budget, upgrade civic services and bolster local demographics, Mitzpe has witnessed repeated rises in property and city taxes and a general hike in the price of real estate including rentals and purchases of both business and residential spaces.

Conclusions

The current study surveyed a six-decade period in which the peripheral Negev desert town of Mitzpe Ramon evolved from a neglected backwater to a reputable ecotourism destination. Evaluating this process from a perspective of CBT (Cole, 2006) and empowerment (Scheyvens, 1999) with regard to different local communities provided insight into the role tourism development assumes in the evolution of the town.

In the three decades following its establishment in 1956, development in Mitzpe Ramon was sporadically underwritten by the government and had a limited connection to tourism. From the 1980s to the early 2000s, the town went through a process of tourism development that, per Butler's fundamental model of the evolution of a tourist area (1980), reaches a sensitive phase, so that the town is currently on the verge of shifting from the (local) "involvement" stage to the (external) "development" stage. Tourism development in Mitzpe Ramon transpired on a grassroots scale and involved individual entrepreneurs who had purposefully vacated the crowded national urban centers in favor of the open and sparsely populated Negev desert. Yet, except for the secular middle-class newcomers, the tourism ventures initiated during this second phase excluded the other local communities, namely the national–religious middle-class gar'in, the long-standing working-class Mizrachim and the Negev Highland Bedouin. As such, the economic impact resulting from the development of the local tourism industry was modest when measured per the number of households that directly benefitted from the town becoming a desert ecotourism destination. Present plans for significant increase in tourism volumes appear to provide advantages to external commercial forces at the expense of local entrepreneurs. Thus, we argue that at present and looking to the future, the process of tourism development in Mitzpe Ramon cannot be regarded as a force that provides even levels of economic empowerment to assorted resident communities.

Disempowerment is also evident in the context of local social cohesion and communal harmony. The influx of liberal–secular and national–religious middle-class communities to an essentially homogenous town of working-class "reluctant pioneers" generated resentment and ensuing political conflict. This process became particularly manifest after the 2013 municipal elections and the subsequent dismissal of the long-standing Mizrachi community from their positions of power. Plans for further commercialization and substantial growth in tourism volume might promote social unrest by adding to the town demographic a new disadvantaged community of temporary low-paid hospitality workers. It might also lead to discord between the traditional-minded residents, chief among them the national religious community, and the secular commercial tourism investors.

It is worth mentioning that regional tourism development is designed and promoted by central government agencies who dutifully adhere to national policy agendas. The data indicate that government strategies and procedures essentially ignore the misgivings of local tourism entrepreneurs who oppose this manner of development. Moreover, it should be noted that the unrecognized Bedouin tribespeople are continually resentful of their restricted involvement in the development of the region. Accordingly, it appears that local tourism development does not engender political empowerment when measured in terms of local access to and influence over decision making processes. Rather, this practice seems to serve the newcomers who acquired political authority in the name of progress, whereas the veteran working-class residents were removed from political power and the Bedouin remain (as they have been for years) powerless.

Unlike the questionable impact on the economic, social and political aspects of empowerment, the development trajectory described above seems to enhance psychological empowerment, whereupon there is a boost in the self-esteem of miscellaneous residents due to the outside recognition of the uniqueness and value of their culture and natural resources. Specifically, we suggest that the association of Mitzpe Ramon with environmentally conscious, nature-based tourism enhances the town image within the symbolic field of class-related Israeli culture. Indeed, Jewish-Israeli mythologies behold the desert as a symbolic site of death *and* rebirth and in the present-day context of the middle-class Israeli imperative, "oppositional and complementary" (Zerubavel, 2008, p. 214) interpretations of the desert continue to influence the contemporary Israeli psyche.

While in the early decades of the state, the desert served the country as the "actual and metaphoric garbage can of Israel" (Feige, 2009, p. 30), its emergence as an attractive place to visit is a significant improvement for the town image. Previous affiliation of Mitzpe with working-class (and even underclass) marginality is being altered as the town rebrands itself as a destination for educated middle-class visitors to experience alternative tourism and personal fulfillment. However, massive development and further commercialization of the tourism industry might also challenge the ability of Mitzpe to endure as a remote "counter-space" (Zerubavel, 2008, p. 211) from which individuals may draw inspiration and revelation.

The contribution of this study to the literature on CBT is both methodological and conceptual. First, the methodological approach to ethnohistory employed the perspective and everyday life concerns of locals (emic) to consider the development processes across time. This framework complies with the premise of CBT that participation and the well-being of locals are key measures of sustainable development. Also, the triangulation of historic record with multifarious ethnographic data furnishes the analysis with internal validity.

Second, the study supports the perspective of CBT that local communities cannot be addressed as a single, unified category since they normally comprise subgroups with different attitudes, interests and power positioning (Clausen & Gyimothy, 2016; Iorio & Corsale, 2014; Lew, 1989). In this regard, the analysis demonstrates that while some local communities may engage in and benefit from tourism development, others are disinterested and/or excluded from this process. Third, the current study complies with the notion of empowerment as a relative multidimensional construct (Scheyvens, 1999). The Mitzpe Ramon case study indicates that while some aspects of empowerment were reinforced by tourism development, others deteriorated.

In addition, the study supports the notion of sustainability as a dynamic condition that may change across phases and planes of tourism development (Saarinen, 2006). The analysis points to an undesirable progression whereupon aspects of resident empowerment evident during the localized stage of tourism development may recede upon the onset of substantial commercialization and dramatic increases in tourism volume. Finally, the transformation of political power in Mitzpe Ramon from veteran Mizrachi working-class residents to a coalition of Ashkenazi dominated, middle-class liberal–secular and national–religious newcomers parallel current modifications in the power relations within Israeli society, particularly the broadening influence of the national–religious sector. However, a discussion of the implications this interesting process holds for Israeli society is beyond the scope of the current paper that focusses on the diverse benefits and costs of sustainable tourism development.

Acknowledgments

The authors wish to thank the Israel Ministry of Science and Technology for their support during the research and writing of this article.

Disclosure statement

No potential conflict of interest was reported by the authors.

References

Adler, C. (1999). *Rationale for the award TO DO! 99 Contest socially responsible tourism.* Retrieved from http\\:www.studienkreis.org.

Ashley, C., Roe, D., & Goodwin, H. (2001). Pro-poor tourism strategies: making tourism work for the poor – a review of experience (Pro-Poor Tourism Rep. 1). London: Overseas Development Institute (ODI); International Institute for Environment and Development (IIED), and the Centre for Responsible Tourism (CRT). Retrieved from https://www.odi.org/sites/odi.org.uk

Ben-Zadok, E. (1993). Oriental Jews in the development towns: Ethnicity, economic development, budget, and politics. In E. Ben-Zadok (Ed.), *Local communities and the Israeli polity: Conflict of values and interests* (pp. 91–122). Albany, NY: State University of New York Press.

Ben-Porat, G., Levi, Y., Mizrahi, S., Naor, A., & Tzfadia, E. (Eds.). (2008). *Israel since 1980*. Cambridge: Cambridge University Press.

Bowden, J. (2005). Pro-poor tourism and the Chinese experience. *Asia Pacific Journal of Tourism Research, 10*, 379–398.

Cohen, E. (2008(. Taking distance: Israeli backpackers and their society. In J. Brauch, A. Lipphardt, & A. Nocke (Eds.), *Jewish topographies: Visions of space, traditions of place* (pp. 265–276). London & Aldershot: Ashgate Press.

Clausen, H.B., & Gyimóthy S. (2016). Seizing community participation in sustainable development: Pueblos Mágicos of Mexico. *Journal of Cleaner Production, 111*, 318–326. Retrieved from https://www.sciencedirect.com/science/article/pii/S095965261500089X

Cole, S. (2003). Cultural tourism development in Ngadha, Indonesia. (Unpublished doctoral dissertation). London Metropolitan University, London, UK.

Cole, S. (2006). Information and empowerment: the keys to achieving sustainable tourism. *Journal of Sustainable Tourism, 14*(6), 629–644.

Elishav, R. (n.d.). *Mitzpe Ramon: Xazon ha'ir hakooperativ* [Mitzpe Ramon: Visions of a cooperative city]. (Seminar Thesis). Available from Tel-Aviv University Library.

Feige, M. (2009). *Midbar, shmama*, and garbage can. In A.P. Hare & G. Kressel (Eds.), *The desert experience in Israel: Communities, arts, science, and education in the Negev* (pp. 27–32). Lanham, MD: University Press of America, Inc.

Gutwein, D. (2004). From melting pot to multiculturalism; or the privatization of Israeli identity. In A. Shapira (Ed.), *Israel identity in transition* (pp. 215–232). Westport, CT: Praeger.

Hall, C.M, & Boyd, S. (Eds.). (2005). *Nature based tourism in peripheral areas: Development or disaster*. Clevedon, UK: Channel View Publications.

Harkin, M.E. (2010). Ethnohistory's ethnohistory: Creating a discipline from the ground up. *Social Science History, 34*(2), 113–128.

Hoffmitz, L. (1978). *The other side of Israel*. Tel-Aviv: Nateev Publishing.

Hyde, K.F., & Lawson, R. (2003). The nature of independent travel. *Journal of Travel Research, 42*(1), 13–23.

IMPART - Integrative multilateral planning to advance rock art tourism (Report No. 3-11655). (2016). Israel Ministry of Science and Technology, Italian Ministry of Foreign Affairs.

Iorio, M., & Corsale, A. (2014). Community-based tourism and networking: Viscri, Romania. *Journal of Sustainable Tourism, 22*(2), 234–255

Jones, S. (2005). Community-based ecotourism: The significance of social capital. *Annals of Tourism Research, 32*(2), 303–324.

Kedar, A., Amara, A., & Yiftachel, O. (2018). *Emptied lands: A legal geography of Bedouin rights in the Negev*. Stanford, CA: Stanford University Press.

Klein Halevi, Y. (2013). *Like dreamers: The story of the Israeli paratroopers who reunited Jerusalem and divided a nation*. New York: Harper Collins.

Lew, A. (1989). Authenticity and sense of place in tourism development: Experience of older retail districts. *Journal of Travel Research, 28*(4), 15–22. Retrieved from http://journals.sagepub.com/doi/abs/10.1177/004728758902700403

Marx, E. (2012). The growth of a conception: Nomads and cities. In H. Hazan & E. Hertzog (Eds.), *Serendipity in anthropological research: The nomadic turn* (pp. 31–46). Farhnam, UK: Ashgate Press.

Mazor, E. (1992). Introduction to the Ramon national geological park. *Israel Journal of Earth Sciences, 42*, 103–114.

McGettigan, F., Burns, K., & Candon, F. (2006). Community empowerment through voluntary input: A case study of Kiltimagh Integrated Resource Development (IRD). In M. Smith & M. Robinson (Eds.), *Cultural tourism in a changing world: Politics, participation and (re)presentation* (pp. 143–159). Clevedon: Channel View Publications.

Meir, A. (1997). *As nomadism ends: The Israeli Bedouin of the Negev*. Boulder, CO: Westview Press.

Meir, A., & Marx, E. (2005). Land, towns and planning: The Negev Bedouin and the state of Israel. *Geography Research Forum, 25*, 43–62. Retrieved from http://raphael.geography.ad.bgu.ac.il/ojs/index.php/GRF/article/view/290

Mingay, G. E. (1989). Introduction. In G.E. Mingay (Ed.), *The rural idyll* (pp. 1–7). London: Routledge.

Nielsen, B.S., & Nielsen, K.A. (2016). Citizens' initiatives for democratic nature management and community development: Reflecting on Danish experiences. In H.P. Hansen et al.(Eds.), *Commons, sustainability, democratization: Action research and the basic renewal of society* (pp. 215–247). New York: Routledge.

Noy. C., & Cohen, E. (2005). Introduction: Backpacking as a rite of passage in Israel. In C. Noy & E. Cohen (Eds.), *Israeli backpackers: From tourism to rite of passage* (pp.1–37). Albany, NY: State University of New York Press.

Plieninger, T., Dijks, S., Oteros-Rozas, E., & Bieling, C. (2013). Assessing, mapping and quantifying cultural ecosystem services at community level. *Land Use Policy, 33*, 118–129.

Ram, U. (2000). The promised land of business opportunities: Liberal post-Zionism and the glocal age. In G. Shafir & Y. Peled (Eds.), *The new Israel: Peacemaking and liberalization* (pp. 217–240). Boulder, CO: Westview Press.

Rapport, N. (1998). Coming home to a dream: A study of the Immigrant discourse of 'Anglo-Saxons' in Israel. In N. Rapport & A. Dawson (Eds.), *Migrants of identity: Perceptions of home in a world of movement* (pp. 61–83). Oxford, UK: Berg.

Reichel, A., & Uriely, N. (2003). Sustainable tourism development in the Israeli Negev Desert: An integrative approach. *Journal of Park and Recreation Administration, 21*(4), 14–29.

Reichel, A., Uriely N., & Shani, A. (2008). Ecotourism and simulated attractions: Tourists' attitudes toward integrative sites in a desert area. *Journal of Sustainable Tourism, 16*(1), 23–41.

Reichner, E. (2013). *Davka sham – Sipuram shel anshei hahityashvut haxevratit [There of all places: The story of the social settlers]*. Tel-Aviv: Yedioth Ahronoth and Chemed Books.

Rivelli, S.D. (2016). Travel in Israel, seven days on the road. *Marie Claire*. June 6, 2016. Retrieved from http://www.marieclaire.it/Lifestyle/viaggi/israele-cosa-vedere-dove-andare#6

Rosenhek, Z. (2011). Dynamics of inclusion and exclusion in the Israeli welfare state: State building and political economy. In G. Ben-Porat & B.S. Turner (Eds.), *The contradictions of Israeli citizenship* (pp. 63–86). London: Routledge.

Saarinen, J. (2006). Traditions of sustainability in tourism studies. *Annals of Tourism Research, 33*(4), 1121–1140. Retrieved from https://www.sciencedirect.com/science/article/abs/pii/S0160738306000909

Schmidt, J. (2015). Shifting perceptions: Negotiating space and place in the Israeli desert frontier town of Mitzpe Ramon. In A. Diekmann & M. K. Smith (Eds.), *Ethnic and Minority Communities as Tourist Attractions* (pp. 99–111). Bristol, UK: Channel View Publications.

Shapira, A. (Ed.). (2004). *Israel identity in transition*. Westport, CT: Praeger.

Shields, R. (1991). *Places on the margin: Alternative geographies of modernity*. London: Routledge.

Scheyvens, R. (1999). Ecotourism and the empowerment of local communities. *Tourism Management, 20*, 245–249.

Sofield, T. (2003). *Empowerment for sustainable tourism development*. Oxford: Pergamon.

Tal, A. (2002). *Pollution in a promised land: An environmental history of Israel*. Berkeley, CA: University of California Press.

Taub, G. (2010). *The settlers and the struggle over the meaning of Zionism*. New Haven, CT: Yale University Press.

Tobin, Y. (2009). Phonology as human behavior: Applying theory to clinic. *Asia Pacific Journal of Speech, Language, and Hearing, 12*(2), 115–134. Retrieved from https://www.tandfonline.com/doi/abs/10.1179/136132809805335409

Toll, R. (2016). A luxurious adventure vacation in a place you'd never expect. *The Wall Street Journal*, May 3, 2016. Retrieved from https://www.wsj.com/articles/a-luxurious-adventure-vacation-in-a-place-youd-never-expect-1462306965

Torres, R., & Momsen J.H. (2004). Challenges and potential for linking tourism and agriculture to achieve pro-poor tourism objectives. *Progress in Development Studies, 4*(4), 294–318.

Turner, F.J. (2008). *The significance of the frontier in American history*. London: Penguin Books.

Turner, V. (1982). *The ritual process: Structure and anti-structure*. London: Routledge and Kegan Paul.

Uriely N., Reichel, A., & Shani, A. (2007). Ecological orientation of tourists: An empirical investigation. *Tourism and Hospitality Research, 7*(3/4), 161–175. Retrieved from http://journals.sagepub.com/doi/abs/10.1057/palgrave.thr.6050045

Weaver, D.B. (2005). Comprehensive and minimalist dimensions of ecotourism. *Annals of Tourism Research, 32*(2), 439–455.

Weingrod, A. (1966). *Reluctant pioneers: Village development in Israel*. New York: Cornell University Press.

Yang, J., Ryan, C., & Zhang, L. (2013). Social conflict in communities impacted by tourism. *Tourism Management, 35*, 82–93.

Zameret, Z., Halamish, A., & Meir-Glitzenstein, E. (Eds.). (2009). *Ayarot pitux [Development towns]*. Jerusalem: Yad Ben Zvi Publishers.

Zerubavel, Y. (2008). Desert and settlement: Space metaphors and symbolic landscapes in the Yishuv and early Israeli culture. In J. Brauch, A. Lipphardt, & A. Nocke (Eds.), *Jewish topographies: Visions of space, traditions of place* (pp. 201–222). London and Aldershot: Ashgate Press.

Community involvement and tourism revenue sharing as contributing factors to the UN Sustainable Development Goals in Jozani–Chwaka Bay National Park and Biosphere Reserve, Zanzibar

Florian Carius and Hubert Job

ABSTRACT

Communities surrounding protected areas (PAs) in developing countries bear disproportionate costs of nature conservation, challenging PA managers to apply participatory approaches, deliver benefits and provide alternative livelihoods. This analysis explores in-depth the roles that nature-based tourism can play in this context. Revenues from visitors to Jozani–Chwaka Bay National Park (NP) in Zanzibar are shared equally between regional government and local communities, enabling PA operations, and providing both individual and collective benefits to residents of the larger UNESCO Biosphere Reserve (BR) surrounding the NP. Findings indicate that the structures and processes qualify as good governance. Locals identified many tangible and intangible benefits from the PA to their communities, ranging from direct employment, social capital development to ecosystem services. Nature conservation, mainly through shared tourism revenues, contributes to all 17 United Nations Sustainable Development Goals (SDGs) in multiple ways. Benefit sharing mitigated land use conflicts, increased residents' acceptance of nature conservation and reduced pressures on ecosystems. The study demonstrates the complexity of revenue sharing arrangements necessary for success, and how to harness sustainable tourism, providing benefits beyond the generation of revenues in which community involvement and the institutional interplay between NP and BR governing stakeholders are key factors.

Introduction

In 2015, the UN Sustainable Development Summit was held in New York as a high-level plenary meeting of the General Assembly and adopted the 2030 Agenda for Sustainable Development (UN, 2015). This document contains 17 Sustainable Development Goals (SDGs) and 169 associated targets that are people-centred, transformative, universal and integrated (Table 1). Tourism is explicitly mentioned in three targets related to Goals 8, 12 and 14 but is considered to potentially contribute to all the goals, directly or indirectly (UNWTO, 2015). This study provides evidence for that potential from a case study, proving that the SDGs are capable of reinvigorating

Table 1. Selected tourism-induced contributions from nature conservation and development in Jozani–Chwaka Bay Biosphere Reserve to the SDGs (source: authors).

Goal	Tourism-induced contributions of conservation and development efforts
1) End poverty in all its forms everywhere	Local civil society organisations established loans and savings schemes.
2) End hunger, achieve food security and improved nutrition and promote sustainable agriculture	Forest is a refuge for wildlife mitigating human-wildlife conflicts on farmlands: farmers are compensated for crop-damage caused by wild animals.
3) Ensure healthy lives and promote well-being for all at all ages	CDF facilitated investments in hospital infrastructure and maintenance and purchased medicine, preventing common diseases and epidemics.
4) Ensure inclusive and equitable quality education and promote lifelong learning opportunities for all	CDF facilitated investments in infrastructure, maintenance and equipment of nursery, primary and secondary schools, covering teachers' salaries, tuition fees and uniforms for pupils and college students.
5) Achieve gender equality and empower all women and girls	Women are notably represented in Village Conservation Committees and in NP-related jobs.
6) Ensure availability and sustainable management of water and sanitation for all	CDF improved water supply and access, sanitation and irrigation farming.
7) Ensure access to affordable, reliable, sustainable and modern energy for all	CDF facilitated rural electrification through establishing power lines.
8) Promote sustained, inclusive and sustainable economic growth, full and productive employment and decent work for all	Jozani–Chwaka Bay NP offers direct employment as well as business opportunities improving the standard of living.
9) Build resilient infrastructure, promote inclusive and sustainable industrialization and foster innovation	CDF helped equip a computer/IT centre.
10) Reduce inequality within and among countries	Foreign visitors to Jozani–Chwaka Bay NP bring a flow of financial capital from industrialised to a least developed country.
11) Make cities and human settlements inclusive, safe, resilient and sustainable	CDF invested in road and lorry maintenance, and secured passenger transport.
12) Ensure sustainable consumption and production patterns	Visitor information centre illustrates the ecological footprint of tourism and encourages sustainable consumption.
13) Take urgent action to combat climate change and its impacts	CDF investments contributed to both climate change mitigation – reforestation of forest and mangroves as carbon sinks – and adaptation – planting trees along roads, reducing heating of roads and minimizing soil warming effects.
14) Conserve and sustainably use the oceans, seas and marine resources for sustainable development	Fishing is prohibited in the NP part of Chwaka Bay, restoring the fish stock of this marine ecosystem.
15) Protect, restore and promote sustainable use of terrestrial ecosystems, sustainably manage forests, combat desertification, and halt and reverse land degradation and halt biodiversity loss	CDF financed reforestation and VCCs patrol in community forests, preventing and detecting illegal activities in protected areas.
16) Promote peaceful and inclusive societies for sustainable development, provide access to justice for all and build effective, accountable and inclusive institutions at all levels	Establishment and empowerment of community and government institutions promoting friendly relations and good governance.
17) Strengthen the means of implementation and revitalize the global partnership for sustainable development	The NP Advisory Committee is an important forum for harmonising efforts between government and community representatives.

sustainable development research (Filho et al., 2018). The five guiding themes of the 2030 Agenda set a framework to transform the design of tourism sustainably: *People*: Tourism serves the needs of hosts and visitors. *Planet*: Tourism respects the environment. *Prosperity*: Tourism provides socio-economic benefits. *Peace*: Tourism builds social capital and promotes friendly relations. *Partnership*: Tourism facilitates understanding and cooperation among stakeholders (Monshausen, Tremel, Plüss, Koschwitz, & Lukow, 2016).

The parties to the Convention on Biological Diversity (CBD, 2010) agreed to legally protect further areas as a key response to biodiversity loss. As many eligible sites are inhabited or used by humans, integrative approaches towards nature-based tourism development as a driver of sustainable development need to be emphasised and promoted to make conservation successful (Job & Paesler, 2013). Both National Parks (NP) and Biosphere Reserves (BR) can be tourist destinations. BR aspire to be model regions for sustainable development and are thus "ideal places to test and develop innovative tourism models that benefit local people and maintain distinctive cultures, biodiversity and associated values" (EuroMAB, 2002, p. 10). The tourism workshop at the 4th World Congress on BRs emphasised that "Sustainable Tourism has the potential to be a driving force for the development of the entire Biosphere Reserve [...] sustaining communities and [...] ecosystem services, thus contributing to the achievement of SDGs" (European Wilderness Society, 2016). Emphasising tourism is often considered as a desirable development option and panacea for various problems.

However, biodiversity conservation can only be sustainable if stakeholders are satisfied with its local benefits and their distribution (Salum, 2009). Effective schemes for sharing tourism's benefits for protected areas, are, therefore essential to ensure the long-term support of local communities (Spenceley, Snyman, & Rylance, 2017). Nature-based tourism guidelines recommend the sharing of revenues from PAs with surrounding communities, but guidance is often limited to this basic rationale and does not extend to instructions on the design of revenue sharing schemes. This indicates a need to analyse good practices in order to share experiences with other sites for potential replication. Furthermore, there is little scientific documentation on how shared funds contribute to regional sustainable development. This study analyses and visualises the governance structures and processes involved with tourism revenue sharing as established in a study site with a NP as the core of a larger BR. Furthermore, it examines how communities invest the shared funds and how far the outputs of their investments contribute to the 17 SDGs.

Sustainable tourism to PAs in developing countries

While international tourism is forecasted to grow in the next decades, aviation's greenhouse gas emissions are significantly retarding efforts to mitigate climate change which is heavily affecting Least Developed Countries (LDCs) and Small Island Developing States (SIDS) (Becken & Hay, 2007). Despite this basic challenge for sustainable tourism, it is a development tool associated with hopes for poverty reduction. Tourism provided 7% of exports from LDCs in 2014 (UNWTO, 2015). In consequence, tourism redistributes wealth towards marginalised regions (Manuel-Navarrete, 2016) despite leakages as revenues are repatriated to import relevant goods and services, and the export of profits to foreign companies (Saner, Yiu, & Filadoro, 2015). But, lack of development can preserve authenticity in nature and culture, which attracts tourists, thus creating capital for LDCs (Drammeh, 2015).

NPs conserve the ecological integrity of larger natural areas with distinctive features, including iconic landscapes or flagship species attracting tourists and financing protection. BRs go far beyond mere area protection and strive to be holistic model regions for sustainable development: a harmonious human-nature relationship does not constitute a significant tourist attraction per se (Job, Becken, & Lane, 2017). Synergies between both concepts are theoretically expected, where a NP attracts tourists to the core area of a BR which then serves as mechanism to share the benefits with the residents. They can deliver additional economic benefits such as supply chain opportunities, strengthening long-term regional sustainable economic development (Kraus, Merlin, & Job, 2014).

Community involvement

The complexity of tourism products as a bundle of services and facilities brings opportunities for PAs to engage in relationships with stakeholders (Font, Cochrane, & Tapper, 2004). Local community in this context is defined as "residents living near (and sometimes within) a protected area" (Leung, Spenceley, Hvenegaard, & Buckley, 2018, p. 98).

Sustainable and inclusive tourism development therefore requires consideration of the heterogeneous interests of local communities and their involvement in managerial decision-making as well as representation in governing institutions (Baumgartner, 2008). Local involvement has various benefits: decisions are based on all available knowledge, democratically legitimated and socially accepted. Participation facilitates rational cooperation, strengthens partnerships, increases stakeholder support, builds trust and credibility, promotes equity, pluralism and good governance, helps to resolve conflicts, builds capacities, creates ownership, empowers communities and marginalised groups and therefore makes management more effective (Amer, Ashong, & Tiomoko, 2015). Constitutive requirements of good governance are the rule of law, respect for human rights, prevention of corruption, democracy, cooperation, transparency, accountability, technical capacity and dedication, effectiveness, efficiency and the sustainability of resources (Borrini-Feyerabend et al., 2013; UNWTO, 2013).

Tourism emerged as a mechanism for funding conservation and consequently for the exclusion of residents from the consumptive use of natural resources (Morais et al., 2018). Land qualifies for protective efforts if it still harbours a significant biodiversity (among other reasons) which to an extent implies that PAs were often established on land previously used by residents in a sustainable manner. In several cases, indigenous people were relocated away from PAs in more or less forced ways (Schmidt-Soltau & Brockington, 2007). Communities surrounding PAs bear disproportionate costs of conservation through human-wildlife conflicts (losing crops or livestock to raiding animals or risking their own health) and restrictions in their use of natural resources (MacKenzie et al., 2017). Conservationists aim to offset these costs and build support for parks by "making informed and fair trade-offs between social, economic and ecological costs and benefits within and between stakeholder groups, and between stakeholders and the natural environment, in a way that is satisfactory to most parties" (Swemmer, Grant, Annecke, & Freitag-Ronaldson, 2015, p. 7).

The Secretariat of the CBD (2004) encouraged governments to channel tourism revenues partially towards conservation of protected areas. A number of international nature-based tourism guidelines all recommend the participation of local communities through consultation about managerial decision-making and representation in governing institutions as well as the provision of benefits including the sharing of monetary revenues (Amer et al., 2015, pp. 35ff. & 48f.; Beyer, 2014, p. 68; CBD, 2004; Denman, 2001; Gebhard, Meyer, & Roth, 2007; Leung et al., 2018, pp. 73ff.; Tapper, 2007, pp. 15, 39f. & 75; UNEP & UNWTO, 2005, p. 65; UNWTO, 2013, p. 20). However, guidance mostly does not extend to instructions on the design of revenue sharing schemes where details matter (Spenceley et al., 2017). Only two of the above mentioned guidelines provide even very short examples, both from East Africa: Volcan(oe)s NP and BR in Rwanda where local communities receive 5% of tourism revenues (Leung et al., 2018, p. 11) and Jozani–Chwaka Bay NP and BR in Tanzania where 50% of revenues are shared with local communities (Amer et al., 2015, p. 49).

Tourism revenue sharing with local communities

Prerequisites, governance and management

The income from tourism in PAs originates from various forms of induced, indirect and direct spending by visitors (Mayer & Job, 2014). Public tourism revenue is mostly dependent on visitation frequency, length of stay and spend levels for entrance or similar fees. Generating enough funds to share with residents is challenging for PA authorities: it relies on significant visitation rates (Butzmann & Job, 2017). Levying fees for visiting PAs require legal authority and functional sharing mechanisms necessitate adequate legislation and policies including political will (Tapper, 2007).

Getting consensual agreement about revenue distribution is difficult. Should it go to all parishes adjacent to PAs, to selected surrounding single villages, or to individual households bearing disproportionate conservation costs? Nomadic tribes present special challenges (Archabald & Naughton-Treves, 2001). Ugandan experience shows that collectively shared revenues can reach a wide range of people (Sandbrook & Adams, 2012). However, in Rwanda, membership of local community associations, which involved membership fees determined eligibility for tourism revenue sharing, excluded socio-economically disadvantaged residents (Munanura, Backman, Hallo, & Powell, 2016). Schnegg & Kiaka (2018) report from Namibia that their communities retain only a small part of revenues from community-based tourism: relatively few people are direct beneficiaries. These examples emphasise the importance of institutional arrangements and access rights (Lapeyre, 2011). Schemes should be acknowledged in legal documents, with systematic guidelines about how they work, to provide reliability for both sides and reduce vulnerability to policy shifts or PA management staff changes which can be frequent in many African countries (Archabald & Naughton-Treves, 2001).

A key issue is always the percentage of tourism revenue to be shared. Examples range from 5% in Rwanda (Munanura et al., 2016), 7.5% in mainland Tanzania (Emerton, Bishop, & Thomas, 2006), 12–20% in Uganda to 10–25% in Kenya. However, the participatory sharing *process* can matter more than the monetary benefits (Archabald & Naughton-Treves, 2001).

Good management practices also require that bureaucracy is minimised, particularly where key local stakeholders are illiterate. And, as schemes can be vulnerable to corruption, good governance is crucial (Snyman, 2017). Tangible local benefits must be identified and communicated (Spenceley et al., 2017). It is therefore recommended to record the revenue and disburse it openly under public supervision.

Revenues should be hypothecated so that they are restricted to specific types of use rather than flowing into the treasury (UNEP & UNWTO, 2005). In Uganda, an Advisory Committee with governmental, civil society organisation (CSO) and community stakeholders selects and monitors projects and reports to the Uganda Wildlife Authority. At community level, a Parish Park Committee composed of village representatives proposes projects and supervises a Project Implementation Committee (Archabald & Naughton-Treves, 2001). In Rwanda, a Revenue Sharing Committee comprising government and CSO representatives calls for proposals and selects projects for funding (Munanura et al., 2016).

In Uganda, proposed projects are submitted with a project description, timeline for implementation, budget and endorsement of the local government. Eligibility criteria comprise budget restrictions, consistency with conservation objectives, sustainability and feasibility. Qualified projects were then prioritised based on the PAs impact on each community, their support for conservation and their in-kind contribution to the proposed project (Archabald & Naughton-Treves, 2001). In Rwanda, projects were selected based on proximity to the NP, impact on conservation and livelihoods and applicant's financial contributions (Munanura et al., 2016).

Basically, revenue received could be invested in development activities supportive of any or none of the 17 SDGs (Engels, Becken, Job, & Lane, 2017). In Rwanda, tourism revenue was invested in a motorcycle competition (Munanura et al., 2016). In Zimbabwe, local demands to build a beer hall were approved. In Uganda, sustainability is assessed when selecting the proposal as mentioned in the preceding section. Based on existing examples, typical investments are dedicated to capacity building or infrastructure development in the fields of agriculture, education, health, waste, water, sanitation, telecommunication and transport (Archabald & Naughton-Treves, 2001).

Monitoring and evaluating implementation

Monitoring is carried out at different scales by different bodies. An effectiveness assessment of Rwandan Nyungwe NP concluded that tourism revenue sharing improved the quality of life (Kamuzinzi, Shukla, & Ndabaga, 2015). The question of who finally benefits from the investments needs careful attention. Well-founded concerns were expressed in Uganda, where the community's upper classes largely benefited. Poor people, in contrast, could not afford to send their children to the built schools: they had to protect their farmland from crop-raiding wild animals or could not pay for medical treatment in the new facilities (Archabald & Naughton-Treves, 2001).

Experience shows the importance of labelling the investments as facilitated through revenue sharing and to provide information during outreach meetings, to make the provided benefits visible and tangible for the beneficiaries (Makame & Boon, 2008). Successful tourism revenue sharing can shift attitudes toward conservation. In many cases, conservation costs were considered higher than the shared benefits (Ahebwa, van der Duim, & Sandbrook, 2012; Emerton et al., 2006; Tumusiime & Vedeld, 2012, 2015). Although the benefit per capita might sometimes be negligible, it can have a collective effect, such as social sanctioning of anti-conservation behaviour or even active support of the area's protection. Once communities know of the opportunities and expect benefits, failure to generate revenues will worsen relations between stakeholders. Moreover, community-level benefits do not compensate individuals for conservation. However, accurately directing tourism revenue to individual households affected by human-wildlife conflicts is a massive administrative challenge: loss claims might be exaggerated to increase compensation (Archabald & Naughton-Treves, 2001).

Rwanda's tourism revenue sharing scheme was criticised as having little impact on conservation and because it prioritised community development, not individual livelihood improvements, attributed to political influence on decision-making (Munanura et al., 2016; Spenceley, Habyalimana, Tusabe, & Mariza, 2010). Another Rwandan study found evidence of inadequate funding, incidents of corruption, inadequate human skills, poor communication and elite capture (Kamuzinzi et al., 2015).

An evaluation of Central African NPs in conserving Western gorillas recommended strengthening transboundary cooperation and treating it as one site in terms of benefit-sharing. It revealed that performance-linked benefit-sharing (based on the maintenance of the gorilla stock) optimised societal conservation outcomes and reduced needs for complementary anti-poaching efforts (Mukanjari, Bednar-Friedl, Muchapondwa, & Zikhali, 2013). Serengeti-Ngorongoro NP and BR in Tanzania combine legal government and voluntary private sector funding mechanisms for communities producing different benefits addressing several conservation aspects thus enlarging the overall impact (Emerton et al., 2006; Job & Schmid, 2011).

Lessons from other revenue-sharing programmes by PA authorities as well as private tourism enterprises in sub-Saharan Africa were reviewed by Spenceley et al. (2017, p. 10). They found four key components in favourable tourism revenue-sharing schemes – park representatives must clearly communicate benefits or potential drawbacks to communities in order to foster positive interactions; – incentives should be appropriate to the scale of biodiversity threats; – locals should be involved in decision making regarding the implementation of a distribution system; – sufficient technical and institutional support must be ensured.

This study selects a single case study that seemed promising for good practice in terms of local community participation, analyses governance and management structures and processes and evaluates contribution towards the SDGs. It identified lessons for potential replication at other PAs and inclusion in future guideline publications.

Case study area

Sub-Saharan Africa was selected as the region for the case study, since most LDCs are located there and land use conflicts between wildlife and humans are widespread. Tourism revenue sharing examples examined were found in Botswana, Cameroon, Central African Republic, Democratic Republic of Congo, Ethiopia, Ghana, Kenya, Lesotho, Mocambique, Namibia, Nigeria, Rwanda, Tanzania, Uganda, South Africa, Zambia and Zimbabwe. Several of the 79 BRs in sub-

Saharan Africa are also NPs. Jozani–Chwaka Bay NP and BR was selected as the study site because its tourism revenue sharing activities with local communities are already attested in the literature (Amer et al., 2015, p. 49; Archabald, 2000; Khatib, 2000, p. 180).

The entrance to Jozani–Chwaka Bay NP is located 35 km south-east of Zanzibar Town. The 2,654 km^2 of the Zanzibar archipelago are inhabited by 1.3 million Zanzibari, all Swahili speakers. Zanzibar is effectively a SIDS with per capita GDP even lower than the average for Tanzania's mainland. Half of the population lives below basic needs poverty. Tourism contributes 20% to GDP (SMZ, 2013) and more than 80% to Zanzibar's foreign exchange earnings (AFED, 2016). Tourist arrivals increased from 22,000 in 1985 to more than 300,000 in 2015 mainly from Italy, Germany and Great Britain (Himmel, 2016; OCGS, 2016).

Jozani Forest is the largest remnant of indigenous forest in Zanzibar (DFNR/Zanzibar, 2015, p. 3) and a key biodiversity area of the "Coastal Forests of East Africa" (CEPF, 2005, p. 79), one of 35 biodiversity hotspots of global importance (Conservation International, 2014).

Jozani Forest was sold by the community to an Indian investor in 1920 and purchased by the British colonial government in 1948 who established a Forest Reserve in 1960 (Finnie, 2003). In the 1980s, Jozani–Chwaka Bay was declared a Conservation Area (DFNR/Zanzibar, 2015) and in 1997 listed as tentative World Heritage of Tanzania (UNESCO WHC, 1997). A large-scale Integrated Conservation and Development Project led to the gazetting of Jozani–Chwaka Bay as the only NP in Zanzibar in 2004 covering 5000 ha (DFNR/Zanzibar, 2015, p. 3). In 2016, UNESCO designated the Jozani–Chwaka Bay BR covering 21,274 ha (UNESCO MAB ICC, 2016). It is inhabited by 16,500 people in 10 shehias (wards) surrounding the NP, the smallest administrative unit consisting of one to two villages.

Tourism here began in the 1990s, mainly motivated by opportunities to observe Zanzibar Red Colobus (*Procolobus kirkii*), a monkey species endemic to Zanzibar and Jozani Forest's flagship attraction. It is Zanzibar's leading ecotourist destination (DFNR/Zanzibar, 2015). The lack of large mammals makes the NP less attractive compared to world class nature-based tourism sites in East African savannas but comparable to other forest PAs in Africa.

Methodology

Theoretical elaborations, and the situation in the field led to three research questions:

1. Through which governing structures and processes are tourism revenues from Jozani–Chwaka Bay National Park shared with the communities in the Jozani–Chwaka Bay Biosphere Reserve?
2. How do communities invest these funds?
3. How do their investments contribute to achieving the Sustainable Development Goals?

The following section provides the methodological foundations of the field research addressing these questions, and also define the scope of the empirical part of this study.

In a transdisciplinary approach to explorative development research, a single case study design was determined by the governance, management and sustainability of tourism revenue sharing with local communities in Jozani–Chwaka Bay NP and BR in Zanzibar. Thirteen semi-structured interviews were conducted face-to-face with key informants (IFAD, 2011) from government, community-based CSOs and governance structures at community level in each shehia currently benefitting from tourism revenue sharing. Within those institutions, representative key informants were selected based on their involvement in tourism revenue sharing and executive power. Interviewees included the BR manager from DFNR, the NP manager heading the respective Authority, the director of Jozani Environmental Conservation Association (JECA), the Secretary of the local smallholder farmers' union (UWEMAJO) plus all Secretaries from nine

Village Conservation Committees (VCCs). These 13 informants were sufficient since all parties to the tourism revenue-sharing scheme were covered adequately.

Based on the theoretical foundations, the state of the field and the specifics of the case, questions were developed on governance, management and sustainability. They were put in a logical sequence that helped the discussion flow, starting with factual issues and concluding with opinions. The questions slightly varied for each informant group, and also covered a range of issues surrounding tourism revenue-sharing in order to get a broader understanding. Field observations through reserved participation, informal discussions with various stakeholders and photographic documentation of artefacts at the case study site were made as an unstructured method but suitable to complement and triangulate interview findings.

The field research project was previously discussed with AfriMAB (UNESCO regional MAB network for sub-Saharan Africa), the Tanzania National Government and the Revolutionary Government of Zanzibar (SMZ)'s Department of Forestry and Non-Renewable Natural Resources (DFNR) who facilitated the research permit issued by Zanzibar Research Committee. Field work encompassed August 2016 based at Jozani Forest Station. The Secretary of JECA was appointed as field assistant for the research project: the 32 years old fisherman from Chwaka had a diploma in teaching, was fluent in Swahili and English and voluntarily served as Secretary for JECA for one year. He organised and translated the interviews with the VCCs and was paid a daily allowance.

No restriction was encountered by any party regarding data collection. All key informants were most willing to share information and made themselves available for interview. No allowances were involved with the four interviews conducted with informants from the government and CSOs. The VCC Secretaries were paid an interview fee of 5,000 TZS (equalled 2,28 US$) which was agreed in advance with JECA. Interviews with the two representatives from government and the CSOs were conducted in English without translation; the remaining interviews with the VCC Secretaries were conducted with Swahili/English translation by the field assistant. Following the interviews, field observations were guided by the interviewees to verify and photographically document the information provided (Carius, 2016). On average, interviews took 2 hours followed by field observations.

Each interview was transcribed based on the audio records and hand written notes. Qualitative data from interviews and field observations were analysed through qualitative content analysis by the interviewing researcher. In a first analytical step, data were explored through a brief check, also looking for serendipity. Secondly, understanding of individual ambiguous aspects was verified with the respective informants or the Field Assistant. The third step required the in-depth dissection of content informant by informant. The structuring content analysis aims to filter a particular structure out of the data through application of deductive categories. Those categories are related to the main research questions and based on theoretical reflections. Categories were derived from the questionnaire and further differentiated according to the topics thus constituting a system of codes and subcodes. Descriptive codes summarise key concepts and aspects deducted before or during the first stage of data collection, saving time and proving to be appropriate. All sections of content suiting a particular code or subcode were coloured and systematically extracted from the material. The subsequent fourth step involved a contextual synopsis of content across informants, correlating data from different sources, and identifying consistent and inconsistent information (Mayring, 2015; Saldaña, 2009; USAID, 1996).

The methods applied are potentially susceptible to different biases occurring during data generation and processing related to the sources, intermediaries and the researching author (as interviewer and analyst) (IFAD, 2011):

Regarding sources, the interviewed representatives from governmental and non-governmental organisations as well as the VCC Secretaries proved to be suitable key informants due to their involvement in tourism revenue-sharing. However, VCC Treasurers would have been valuable additional key informants when it came to accounting. In several cases, VCC Secretaries brought the relevant documents or the VCC Treasurers along to the interview. If they took office only recently, some were accompanied by their predecessor to fill potential knowledge gaps.

Intermediaries were first the representatives of the NP Authority and JECA in the initial meeting where the key informants and field assistant were selected but the choices proved useful. The fact that the field assistant served as JECA Secretary and in this capacity deals with the VCC Secretaries on the subject matter was not perceived as detrimental to the interviewees' willingness to frankly share information on a meaningful extent. This potential conflict of interest did also not materialise because he was relatively new in this position and seemed younger than all interviewees. The scale of information lost during translations between Swahili and English is difficult to estimate and it needs to be noted that he had no professional experience as interpreter. Swahili was his mother tongue and his English skills proved to be totally sufficient.

Potential biases related to the investigating author were mitigated as far as possible but cannot be certainly avoided, and neither can his evidence assessment. The elaborated questionnaire proved to be useful for understanding the situation and reaching the research objectives. The semi-structured interviews were focused so that the answers of different key informants to the same topic can be correlated. The interviewer felt familiar with the issues and contexts, sought clarification in a conversational manner and probed for information spontaneously. Nevertheless, he adopted a neutral attitude, thus acknowledging the exploratory nature of this study and not to only pick up information that confirmed preconceived ideas. Although no hesitation by interviewees to share information with the interviewer was noted, the effect of cultural difference cannot be estimated.

Results

This section is written in great detail to illustrate the vital complexities of local revenue sharing.

Generating tourism revenues

Jozani Forest Station is currently the only developed entry point for NP visitors. Although easy to reach by public transport, most tourists hire local tour operators for day excursions which usually include a stopover for a few hours at the NP. Visitors are received only at the reception and information centre jointly staffed by government and community from 7:30 to 17:00 daily. Non-residents pay an entrance fee of either 20,000 TZS or 10 US$ while residents pay half that price and children only pay half of those prices. Fees were introduced by the government in the early 1990s. Payment permits access to Jozani Forest as the central part of the NP as well as to Pete Community Mangrove Boardwalk located outside the NP (but within the BR) (Figure 3) accompanied by a guide.

Collection seasons last 6 months after which all shareholders such as DFNR, Jozani–Chwaka Bay NP, JECA, UWEMAJO and VCC of Pete gather together to adopt the collection report. No party has ever noticed leakages of revenues collected. During the fiscal year July 2017–June 2018, 59,206 visitors were recorded for Jozani–Chwaka Bay NP generating 1,019,927,230 TZS of revenue (equivalent to 447,111 US$[1]) sourced only from entrance fees. Both numbers increased significantly during the last three years (Figure 1). Marketing is rather scanty: rising visitor numbers are explained by the popularity of traveller IT-platforms rating the attraction highly.

Sharing of tourism revenues

From the early 1990s until 1997, all revenues from visitors were collected by the SMZ Treasury (Makame & Boon, 2008). Tourism revenue sharing was encouraged by law in 1996, agreed with the government on request of communities in 1997 and fully operationalised in 2001 (Jambiya, Puri, & Risby, 2004). A baseline study conducted in the early years of the NP (Jumah, Othman, Said, & Hakim, n.d.) identified farmers as the most affected group by conservation, and consequently, their portion was raised. After some changes over time, since 2009, 50% of total

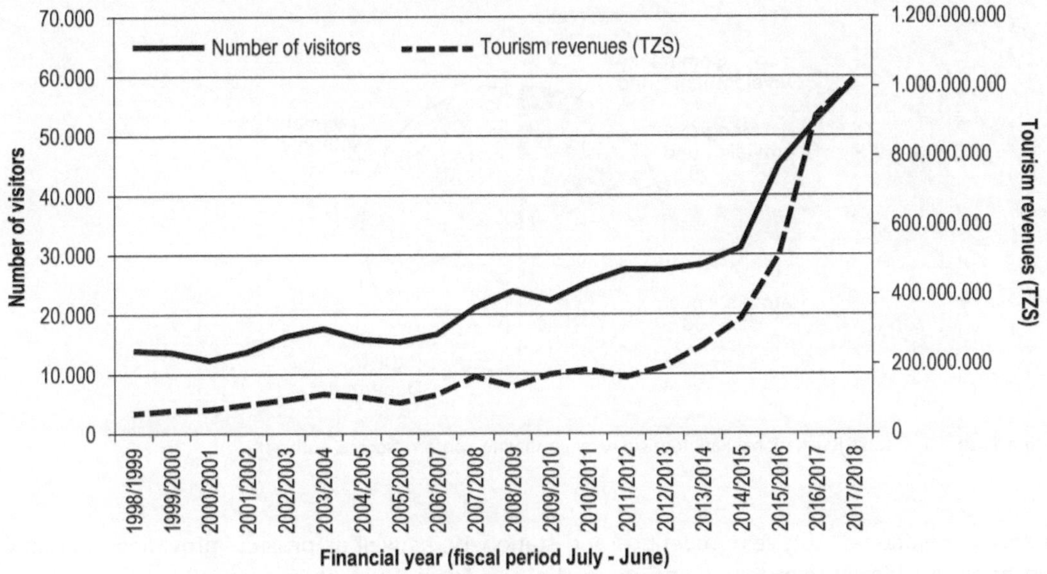

Figure 1. Number of visitors and revenues in Tanzania Shillings (TZS) collected for Jozani–Chwaka Bay National Park over the past 20 years (compilation: authors; data: DFNR/SMZ).

revenues remain with the government (32% Jozani–Chwaka Bay NP Authority (JCB-NPA) and 18% DFNR) and 50% with CSOs (30% UWEMAJO, 4% JECA, 8% Community Development Fund (CDF), 8% VCC of Pete for Mangrove Boardwalk Fund), as agreed in a memorandum of understanding (Figure 2).

In subsequent public tourism revenue sharing ceremonies conducted every 6 months, the Zanzibar Minister of Agriculture, Natural Resources, Livestock and Fisheries officially declares the results and hands over cheques to shareholders. Participating local journalists report the news. All shareholders gather monthly at the Jozani–Chwaka Bay NP Advisory Committee meetings where each party reports its relevant activities. Although there is no formal general reporting duty concerning the usage of their share of the tourism revenues, this information is usually provided on this occasion.

Jozani–Chwaka Bay NP Authority is a governmental institution that belongs to DFNR. Jozani–Chwaka Bay NP has 57 staff (including 6 volunteers and 5 community workers employed by VCC of Pete – the remaining majority paid by DFNR separately, i.e. not from tourism revenue) with more than 90% recruited from the Jozani–Chwaka Bay BR communities. The revenues from the 32% of revenues for Jozani–Chwaka Bay NP are used for the operation of the NP according to the General Management Plan (Finnie, 2003) and the Annual Work Plan.

DFNR is subordinated to the Zanzibar Ministry of Agriculture, Natural Resources, Livestock and Fisheries and superintending Jozani–Chwaka Bay NP Authority. DFNR has 736 staff. The 18% DFNR revenue share feeding the DFNR budget directly for its own use and not through the SMZ Treasury. It pays for travel costs, allowances/field meeting, the conservation section's office stationery/equipment costs and for firefighting.

The 99 owners of small farms around Jozani–Chwaka Bay NP are automatically members of their union, Umoja wa Wenye Mashamba Jozani (UWEMAJO). Membership is based on the location of the farmland (including five shehias, thus excluding seven shehias which otherwise benefit from CDF). UWEMAJO presents its tourism revenues to its members at the semi-annual General Meeting where previous and new investments are discussed and decided. From the UWEMAJO portion, 20% are deducted as membership fee and kept as institutional budget. This

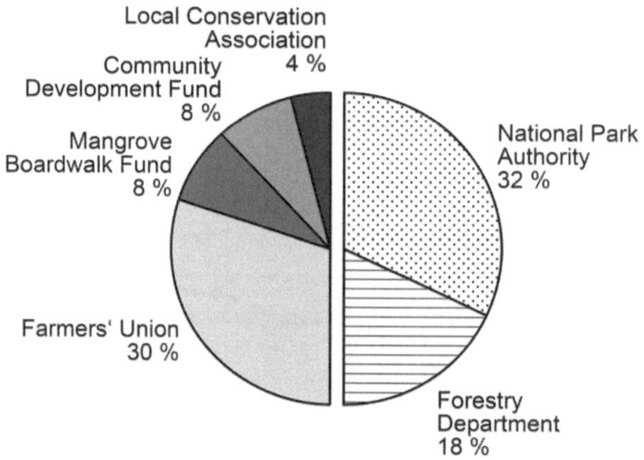

Figure 2. Jozani–Chwaka Bay National Park tourism revenue sharing portions (source: authors).

finances operational costs (e.g. meetings and stationery) as well as projects providing alternative livelihoods to local communities, including the Sea Turtle Sanctuary, ecotours and greenhouse farming. UWEMAJO distributes the remaining 80% to members in cash. The individual amount is quite small in view of the twice yearly allocation and the number of people it needs to feed (big family size). Farmlands are categorized in three sizes with owners of larger plots gaining proportional extra shares.

Individual residents of nine shehias surrounding Jozani–Chwaka Bay NP are constitutionally members of JECA, represented by their respective VCC. The 4% of tourism revenue sharing for JECA covers operational costs of the organisation (stationery, building maintenance). JECA Leadership Committee decides the usage of these funds, reporting to their Executive Committee quarterly. JECA operates a Community Gift Shop, another source of tourism-related funding.

The 8% of tourism revenue for CDF managed through JECA is distributed 10% to each of nine benefitting shehias (Figure 3) and 10% remaining with JECA for overheads. These include monitoring visits as well as training for VCCs. JECA has an Executive Committee composed of two representatives of each VCC of the nine shehias who are informed about tourism revenue sharing. CDF is divided automatically into the respective parts. At their annual general meeting, JECA reports about collected and shared revenues. CDF is allocated via bank cheque to treasurer of each VCC upon their request accompanied by a project proposal approved by JECA. JECA guides VCCs to use <40% of CDF for conservation and >60% for community development. VCCs report to JECA on CDF use and JECA reports to DFNR. There is a donation box for CDF at the tourist reception desk which earns extra 200,000–300,000 TZS per year. This is dedicated to conservation only and shared similarly 10% for each VCC of the nine shehias and 10% for JECA.

In coordination with DFNR, VCCs were constituted in each shehia surrounding Jozani–Chwaka Bay NP through statutes. The 25 to 38 members are elected for several years by the shehia assembly from the entire community and are replaced upon individual appointment by the VCC. Only seven of the nine benefitting shehias are part of the Jozani–Chwaka Bay BR. Michamvi and Bwejuu benefit from CDF because they are adjacent to Chwaka Bay and thus relevant for the conservation of this ecosystem. In contrast, the three western shehias of Bungi, Kikungwi and Jendele are part of the Jozani–Chwaka Bay BR but do not benefit from CDF (Figure 3). These conceptual inconsistencies in the geographies of the BR and the tourism revenue sharing beneficiaries can only be explained by historic reasons linked to either the forest or the bay ecosystems.

The VCC and the Village Development Committee (VDC) usually discuss together potential investments suggested by their members and take a joint decision in coordination with the

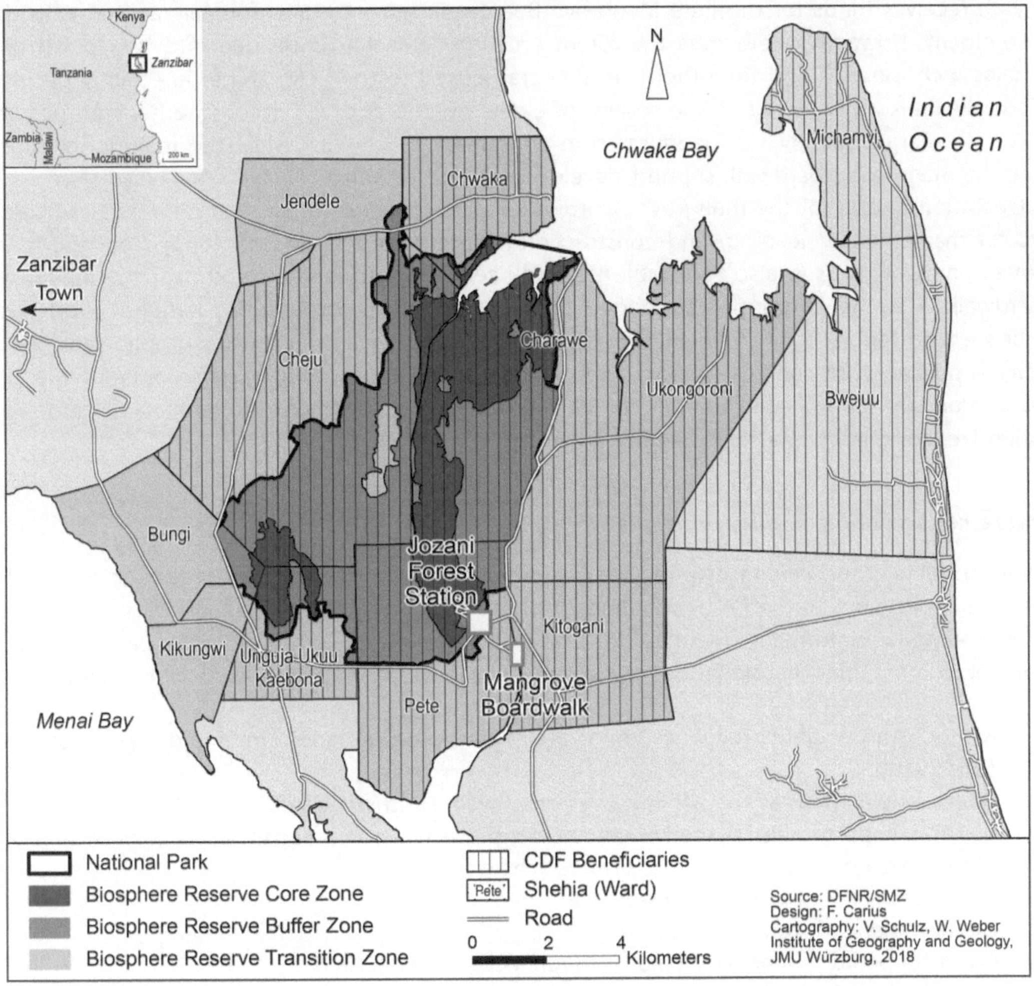

Figure 3. Map of Jozani–Chwaka Bay National Park and Biosphere Reserve at Unguja Island in Zanzibar/Tanzania indicating CDF beneficiaries and the two key points of interest for tourists: Jozani Forest Station and Mangrove Boardwalk.

sheha (chief of shehia). CDF was invested in various conservation and development activities (see Table 1).

It was noted that similar activities were continuously funded over several periods. Furthermore, CDF investments were sometimes completed by governmental investments and vice versa. A delegation of VCC and VDC members monitors project progress on site. After project completion, the VCC reports through the sheha to JECA. The shehia assembly is also informed about the results and can discuss ideas for future projects. Investments are not communicated via signboards as requested by JECA but it is usually clear to all community members that the particular output was facilitated by CDF who are aware that it is derived from tourism revenue sharing of Jozani–Chwaka Bay NP. Poor, women, young and old people were identified as the most advantaged groups by CDF within the communities.

So far, accommodation facilities can only be found in three shehias with beach access while three other shehias plan to have facilities soon. There is no tourism development going on in the other villages although some attractions can be found outside the settlements. Communities do not benefit from tourism beyond Jozani–Chwaka Bay NP tourism revenue sharing, according to their perception.

Pete receives funds for the Pete Mangrove Boardwalk since 1996 as formally agreed with the government. Nowadays Pete gets 8% for the mangrove which is divided into two parts: one remains with the VCC and the other is further transferred to the VDC. Shares are jointly decided in coordination with sheha and the council of elders by VCC and VDC (there are no fixed percentages). The larger part of CDF is invested in conservation activities while the majority of funds from the mangrove boardwalk support development in the shehia.

Operational costs for the mangrove boardwalk include salaries for six staff members (4 guides, two for maintenance/cleaning) and construction/maintenance/reparation. Money from the mangrove boardwalk is used also to establish a small zoo and a fruit sale stand near the mangrove boardwalk. A women's group is supported by supplying ropes for seaweed farming. Funds also settled a land tenure conflict in court. Further investments: include a well to supply water, electricity supply to communities, a foundation stone for the mosque, four classrooms for the primary/secondary school, contributions for individual medical treatment of common diseases and tuition fees for guides to attend training at college.

Nature conservation implications

Several aspects of nature conservation were addressed through the interviews about positive and negative implications of the PA designation in order to consider their correlation with tourism revenue sharing. Key informants identified diverse positive effects of Jozani–Chwaka Bay NP on surrounding communities that go far beyond ecosystem services (listed below without ranking):

1. NP tourism generates revenues which are shared promoting community development and conservation.
2. Direct employment at the NP improves employee living standards.
3. Business opportunities at the NP entrance and at smaller ecotourism attractions nearby.
4. Introduction of credit, loans and savings schemes.
5. CSO establishment enhanced capacities.
6. Establishment of women's groups.
7. Good cooperation among shehias through VCCs under JECA.
8. Friendly relations and informal assistance between government and communities.
9. Adoption of land use plans and Community Forest Management Agreements for each shehia or private wood lots promoting sustainable use maintaining ecosystem services (catering for supply of natural resources which are prohibited to harvest inside NP).
10. Visiting healing sites, licensed harvesting of honey from wild bees or medical plants inside NP.
11. Training on agriculture and beekeeping.
12. Environmental education in NP inspires residents about conservation within own shehia and increases awareness levels.
13. The forest with its particular precipitation regime is a source of rainfall increasing agricultural yield.
14. Refuge for wildlife reducing human-wildlife conflicts outside NP.
15. Recreational enjoyment of biodiversity and good environment in general.

In contrast, the following negative effects on residents of nature conservation through the area protection of Jozani–Chwaka Bay were registered (also without ranking):

1. Restrictions in use of natural resources inside the PA for domestic and commercial purposes, including firewood collection, timber extraction, charcoal making, limestone mining, livestock grazing, fishing, hunting, roof thatching material and medicinal plant collection.

2. Farmers are not allowed to harm or scare crop-raiding animals on farmland, particularly Zanzibar Red Colobus. For the same reason, use of watchdogs for livestock is restricted.
3. Land tenure changes: Loss of land then included in NP, although new land was allocated as compensation.
4. Possible fires spreading from Jozani Forest to adjacent areas.

These negative effects could be mitigated through proper implementation of existing sustainable land use policies outside Jozani–Chwaka Bay NP which would reduce pressure on this PA as resource supply could come from other areas within Jozani–Chwaka Bay BR such as wood lots or Community Forest Management Agreements. There are also some exceptions for environmental friendly resource use where licences are issued according to NP General Management Plan or NP rangers (forest guards) accompany individuals seeking benefits from particular ecosystem services such as healing sites or subsistence honey harvesting from wild bees (the main cause of destructive fires when smoking bees out of their hives to ease harvesting). All key informants stated that the positive dominate over negative effects of Jozani–Chwaka Bay NP on surrounding communities.

Analysis and discussion of results

Governance

The complex structures described above and the complexities of tourism revenue sharing of Jozani–Chwaka Bay NP (Figure 4) qualify as good governance because: Basic law rules, press freedom is fostered, corruption and leakages are prevented, participation and elections of representatives promote democracy, local communities are empowered through involvement, operations are transparent, accountable, functional and decentralised towards stakeholders.

The tourism revenue sharing scheme is operated by institutions whose existence can be linked to the NP but whose mandate reaches beyond tourism revenue sharing so they are legitimated for, and dedicated to, multi-purposes, adding to their institutional sustainability. The patron of the local smallholder farmers' union (UWEMAJO) stated that community involvement in decision-making processes and structures is actually more important than the budget to be shared. Inconsistencies were noted in the geographical scope where out of 12 shehias in the area, the BR only covers ten of them while nine other benefit from CDF (Figure 3) and UWEMAJO only represents five shehias.

Management

Given the very special kind of NP tourism with day-trippers staying only for two hours on average and hardly intruding into the PA[2], the carrying capacity of the ecosystem does not seem exhausted yet (Salerno et al., 2013). Nevertheless, the 1997 Tourism Plan for Jozani–Chwaka Bay Conservation Area fixed the carrying capacity at 25,000 annual visitors (exceeded since 2010) partly to maintain the quality of the visitor's experience (Finnie, 2003). This is also a serious issue with perceptible adverse effects on tourism. To best get the nature experience, and for visitors to learn about sustainable development, it would be desirable to increase the visit duration to over half a day. This could, however, threaten the overall experience and further stress the limits of acceptable change of the ecosystem (Lama & Job, 2014). Visitor numbers are expected to grow further as the destination becomes more popular. Dikgang et al. (2017) suggest that there might be potential to increase entrance fees for international visitors which might slightly decelerate the growing number of visits but probably not total revenues.

NP tourism to Jozani–Chwaka Bay matches all points of the ecotourism definition as travel is responsible and contributes to conservation, well-being of residents and education (TIES, 2015). The development and management of tourism is mainly the task of the Jozani–Chwaka Bay NP

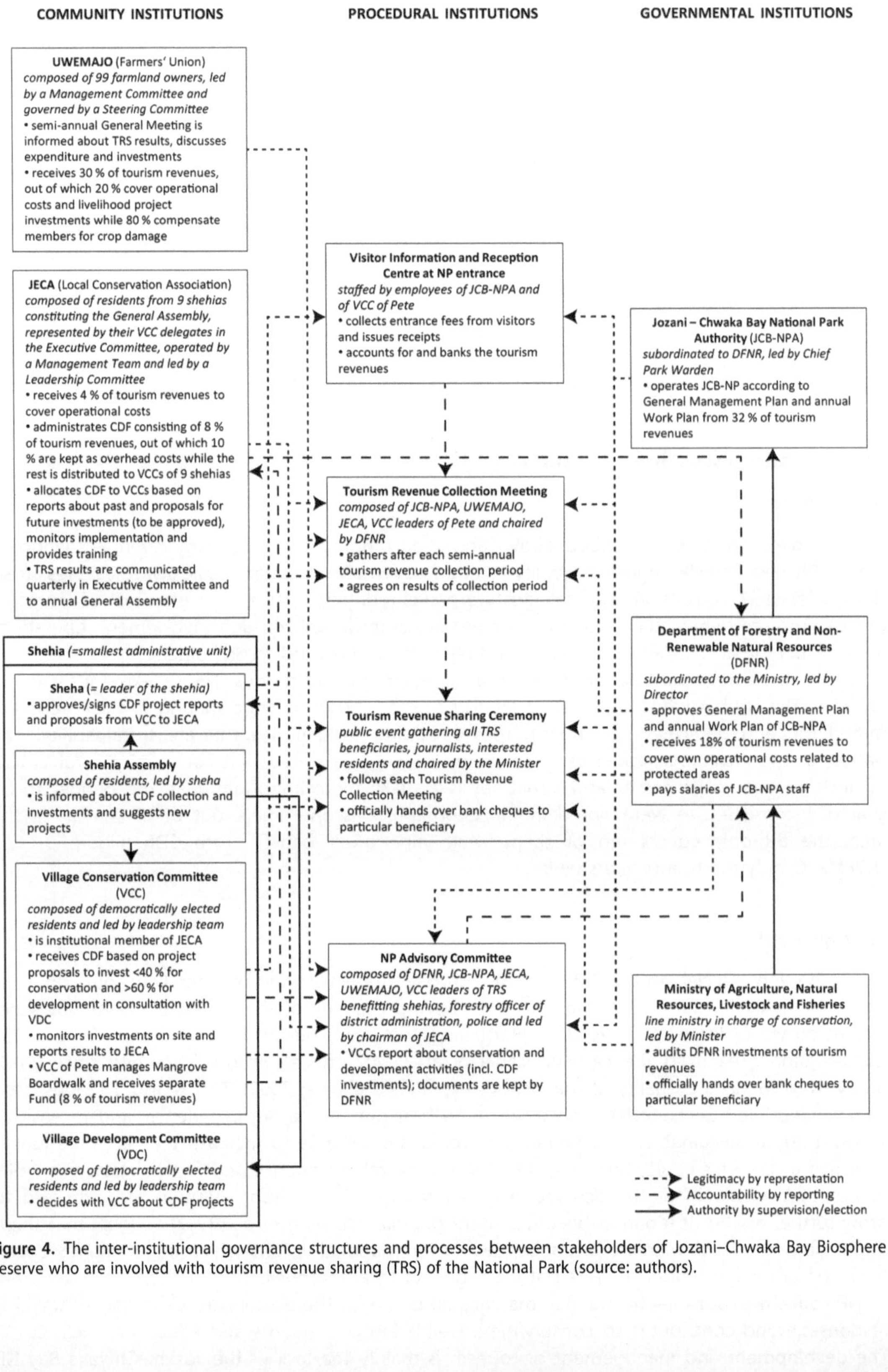

Figure 4. The inter-institutional governance structures and processes between stakeholders of Jozani–Chwaka Bay Biosphere Reserve who are involved with tourism revenue sharing (TRS) of the National Park (source: authors).

Authority. It does not fully qualify as community-based tourism. However, the communities are involved in the broader process of conservation, development and do benefit from tourism. Avoiding tourism in the villages also prevents creating discrepancies between development winners and losers.

Needs for capacity building were documented in various fields, while administrative issues need attention. Although JECA requests the appropriate labelling of CDF investments, this is not done by VCCs yet. Sustainable development efforts facilitated through tourism revenue sharing are not communicated to the visitors. The CDF donation box at the cashier's table is not well filled due to a lack of awareness and appreciation. There is a case for professionalised presentation that could advance donations, and improve the visitor experience.

Contributions to the sustainable development goals

The generous 50/50% tourism revenue sharing between government and civil society institutions seems unprecedented from an Africa wide perspective. Table 1 indicates how nature conservation and related development efforts in Jozani–Chwaka Bay BR contributed to each of the 17 SDGs, considering their associated 169 targets (UN, 2015). These contributions were enabled by successful nature-based tourism governance using tourism revenue sharing to facilitate investments and better acceptance of nature conservation measures, thus seeking long term success.

Tourism's potential to contribute to all SDGs can be clearly evidenced. Money can fund any development and it is therefore not surprising that tourism through revenue sharing can contribute to all 17 SDGs. It is, however, really remarkable that in this single case, local civil society and government institutions jointly manage contributions to all SDGs without even knowing them. The capacity of local stakeholders to deliver SDG contributions should not be underestimated. It also proves that the SDGs, as elaborated in a broad global consultation process resulting in an international standard, actually match the development priorities of local communities and are very relevant at local level, especially in the PA context (Job et al., 2017).

Lessons learned

Tourism revenue sharing at Jozani–Chwaka Bay NP and BR provides many lessons about the theory and praxis of sustainable regional development in a PA context. Success is due to 12 key factors:

1. There are few staff changes due to the island character and political autonomy of Zanzibar. Rotation within governmental institutions enables shifting positions but working on similar issues. Good governance structures and processes provide an anti-corruption framework.
2. Over 90% of staff of the NP Authority is recruited from the BR communities. This reduces the usual government vs. community dichotomy, overcame previous conflicts and prepared the ground for joint efforts.
3. Community involvement in governance and management is time consuming but pays off. Participation might be more important than the actual budget of tourism revenue sharing.
4. A baseline study assessing the situation was conducted by the government to set up the tourism revenue sharing scheme, identifying beneficiaries and calculating their respective share. That scientific evidence was transparent, accepted by local stakeholders, reducing the vulnerability to jealousy.
5. The tourism revenue sharing scheme is operated by institutions with a mandate beyond tourism revenue sharing. They are legitimated for and dedicated to multi-purposes, adding to their institutional sustainability.

6. Tourism revenue sharing empowers local communities to invest according to their priorities. No government institution interferes with the investment decisions of civil society institutions.
7. Independent investments increase the ownership of the outputs established. Community investments consider sustainability while maintaining larger government investments and vice versa.
8. Investments are monitored and evaluated by the investors which might not be independent in a scientific manner but encourages reflection on decisions and progress.
9. 50/50% tourism revenue sharing between government and civil society institutions is considered fair and equitable by all beneficiaries. Even though communities appreciate its benefits, the amount of revenue shared is not enough to compensate for the costs of conservation as established by the baseline study and perceived by the beneficiaries.
10. Tourism mobilises significant financial resources which are hypothecated, that is, re-invested in conservation and development of the NP and BR by all beneficiaries. Tourism revenue sharing provides collective benefits for BR communities and individual compensation for most affected villagers fairly distributed.
11. Tourism revenue sharing is a market-based approach of generating revenues without creating donor-recipient dependencies. No development cooperation is required for progress in villages.
12. The ideal relationship between NP and BR produces the expected conceptual synergies where the NP attracts tourists and the BR serves as a larger geographical PA framework for the fair distribution of tangible and intangible resident benefits.

Bearing in mind specific local situations in determining applicability and transferability to other contexts, the above insights can guide practitioners and scholars in designing successful tourism revenue sharing schemes at other sites in Zanzibar, Tanzania, Africa or LDCs elsewhere.

Conclusion

This study revealed that nature-based tourism to Jozani–Chwaka Bay NP is a success story of which locals seem well aware, unlike academia or the international public. Jozani–Chwaka Bay NP and BR can be considered as good practice examples, linking conservation and tourism to catalyse sustainable regional development based on good governance. Tourism was the key leveraging factor that induced a development process in the BR based on nature conservation in the NP which is sustainable in SDG terms. Tourism delivers benefits beyond the generation of revenues and nature conservation delivers benefits beyond ecosystem services.

Research desiderata include options for diversifying and improving the tourism product to reduce crowding effects, assessment of the role of tourism revenue sharing for mitigating human-wildlife conflicts and further reflections on the context of the institutional interplay between overlapping NPs and BRs.

Notes

1. Interbank exchange rate of OANDA currency converter (https://www.oanda.com/currency/converter/) as of 1 July 2018.
2. Based on own field observations, one can estimate that >95% of visitors stay within the less ecologically sensitive 1 km^2 around the entrance of the NP which is generally only accessible on foot.

Acknowledgements

We are grateful to the Revolutionary Government of Zanzibar, especially DFNR and Zanzibar Research Committee, for permitting this research and to the German Academic Exchange Service (DAAD) for the grant received. We thank all key informants for frankly sharing their insights and all facilitators of the field work, in particular Ali Ali Mwinyi, Mgeni Sheha Ali, Tahir Abass Haji and Mzee Khamis Mohammed. Finally, we thank the reviewers for their constructive critiques of an earlier version of this paper.

Disclosure statement

No potential conflict of interest was reported by the authors.

References

AFED (2016). *Tanzania: Tourism seen booming in Zanzibar*. Retrieved from https://africadata.com/blog/2016/11/04/tanzania-tourism-zanzibar-booming/

Ahebwa, W. M., van der Duim, R., & Sandbrook, C. (2012). Tourism revenue sharing policy at Bwindi Impenetrable National Park, Uganda: A policy arrangements approach. *Journal of Sustainable Tourism*, *20*(3), 377–394. doi: 10.1080/09669582.2011.622768

Amer, W., Ashong, S., & Tiomoko, D. (2015). *Management manual for UNESCO biosphere reserves in Africa: A practical guide for managers*. Bonn: German Commission for UNESCO.

Archabald, K. L. (2000). Can revenue sharing save wildlife? A case study of Jozani Chwaka Bay Conservation Area Zanzibar, Tanzania. (Unpublished Masters thesis, Geography). University of Wisconsin, Madison.

Archabald, K. L., & Naughton-Treves, L. (2001). Tourism revenue-sharing around national parks in Western Uganda: Early efforts to identify and reward local communities. *Environmental Conservation*, *28*(2), 135–149. doi:10.1017/S0376892901000145

Baumgartner, C. (2008). *Nachhaltigkeit im Tourismus: Von 10 Jahren Umsetzungsversuchen zu einem Bewertungssystem. [Sustainability in Tourism: From 10 years of implemention attempts toward an evaluation system]* Innsbruck: StudienVerlag.

Becken, S., & Hay, J. E. (2007). *Tourism and climate change: Risks and opportunities*. Clevedon: Channel View Publications.

Beyer, M. (2014). *Tourism planning in development cooperation: A handbook. Challenges—Consulting approaches—Practical examples—Tools*. Eschborn: GIZ.

Borrini-Feyerabend, G., Dudley, N., Jaeger, T., Lassen, B., Pathak Broome, N., Phillips, A., & Sandwith, T. (2013). *Governance of protected areas: From understanding to action*. Gland: IUCN.

Butzmann, E., & Job, H. (2017). Developing a typology of sustainable protected area tourism products. *Journal of Sustainable Tourism*, *25*(12), 1736–1755. doi:10.1080/09669582.2016.1206110

Carius, F. (2016). Tourism Revenue Sharing with Local Communities: A Case Study of Jozani—Chwaka Bay National Park and Biosphere Reserve, Zanzibar. (Unpublished Masters thesis, Sustainable Development Cooperation). Technical University of Kaiserslautern, Kaiserslautern.

CBD (2004). *Guidelines on biodiversity and tourism development*. Montreal: Secretariat of the Convention on Biological Diversity.

CBD (2010). *Strategic plan for biodiversity 2011–2020 and the Aichi targets*. Montreal: Secretariat of the Convention on Biological Diversity. Retrieved from https://www.cbd.int/doc/strategic-plan/2011-2020/Aichi-Targets-EN.pdf

CEPF (2005). *Eastern Arc mountains & Coastal forests of Tanzania & Kenya—Ecosystem profile* (2nd ed.). Nairobi: Critical Ecosystem Partnership Fund. Retrieved from http://www.cepf.net/Documents/final.easternarc.ep.pdf

Conservation International (2014). *Hotspots*. Retrieved from http://www.conservation.org/how/pages/hotspots.aspx

Denman, R. (2001). *Guidelines for community-based ecotourism development*. Gland: WWF.

DFNR/Zanzibar. (2015). *Proposed Jozani-Chwaka Bay National Park to be Biosphere Reserve (JCBNP-BR) Zanzibar* [Jozani-Chwaka Bay Biosphere Reserve Nomination Dossier]. Zanzibar: Ministry of Agriculture and Natural Resources—Department of Forestry and Non-Renewable Natural Resources.

Dikgang, J., Muchapondwa, E., & Stage, J. (2017). Securing benefits for local communities from international visitors to the Kgalagadi Transfrontier Park. *Tourism Economics*, *23*(8), 1553–1567. doi:10.1177/1354816617707593

Drammeh, F. Y. (2015). Model for sustainable tourism development for least developed countries (LDCs). *African Journal of Hospitality, Tourism and Leisure*, *4*(2), 1–20.

Emerton, L., Bishop, J., & Thomas, L. (2006). *Sustainable financing of protected areas: A global review of challenges and options*. Gland: IUCN.

Engels, B., Becken, S., Job, H., & Lane, B. (2017). An interview with a protected area insider. *Journal of Sustainable Tourism*, *25*(12), 1895–1898. doi:10.1080/09669582.2017.1380365

EuroMAB (2002, May 24–25). Ecotourism and sustainable development in biosphere reserves: Experiences and prospects. Workshop Summary Report. Quebec City: Canadian Commission for UNESCO. Retrieved from http://unesdoc.unesco.org/images/0012/001277/127757e.pdf

European Wilderness Society. (2016). *Message on sustainable tourism at 4th Congress on Biosphere Reserves*. Retrieved from http://wilderness-society.org/message-on-sustainable-tourism/

Finnie, D. (2003). *Jozani—Chwaka Bay National Park—General Management Plan*. Zanzibar: Department of Commercial Crops, Fruits and Forestry.

Filho, W. L., Azeiteiro, U., Alves, F., Pace, P., Mifsud, M., Brandli, L., … Disterheft, A. (2018). Reinvigorating the sustainable development research agenda: The role of the sustainable development goals (SDG). *International Journal of Sustainable Development & World Ecology*, *25*(2), 131–142. doi:10.1080/13504509.2017.1342103

Font, X., Cochrane, J., & Tapper, R. (2004). *Pay per nature view: Understanding tourism revenues for effective management plans*. Leeds: Leeds Metropolitan University.

Gebhard, K., Meyer, M., & Roth, S. (2007). *Sustainable tourism management planning in biosphere reserves—A methodology guide*. Bonn: ETE. Retrieved from https://www.oete.de/images/dokumente/projekt_gef/ETE_Sustainable_Tourism_Management_Planning_2007.pdf

Himmel, S. (2016). Challenges and prospects for sustainable tourism in small island developing states—the case of Zanzibar. (Unpublished Masters thesis, International Studies of Leisure and Tourism). Bremen City University of Applied Sciences, Bremen.

IFAD. (2011). Key Informant Interviews. In IFAD Asia Pacific Division (Ed.), *M&E/KM Toolkit* (pp. 313–320). Rome: International Fund for Agricultural Development. Retrieved from http://asia.ifad.org/documents/627927/327cf919-8515-44ed-b40c-e1a32ad7e7e7

Jambiya, G., Puri, J., & Risby, L. A. (2004). *The nature and role of local benefits in GEF programme areas; case study Tanzania—Zanzibar: Jozani-Chwaka Bay Conservation Project*. Washington, DC: GEF. Retrieved from http://www.easternarc.or.tz/groups/webcontent/documents/pdf/Local_Benefits_Case_Study_Tanzania_Biodiversity.pdf

Job, H., Becken, S., & Lane, B. (2017). Protected areas in a neoliberal world and the role of tourism in supporting conservation and sustainable development: An assessment of strategic planning, zoning, impact monitoring, and tourism management at natural World Heritage Sites. *Journal of Sustainable Tourism*, *25*(12), 1697–1718. doi:10.1080/09669582.2017.1377432

Job, H., & Paesler, F. (2013). Links between nature-based tourism, protected areas, poverty alleviation and crises—The example of Wasini Island (Kenya). *Journal of Outdoor Recreation and Tourism*, *1–2*, 18–28. doi:10.1016/j.jort.2013.04.004

Job, H., & Schmid, L. (2011). 'Serengeti darf nicht sterben!'—50 Jahre integriertes Nationalparkmanagement in Tansania. ['Serengeti shall not die!'—50 years of integrated national park management in Tanzania]. *Natur und Landschaft*, *86*(2), 52–58. doi:10.17433/2.2011.50153080

Jumah, S. M., Othman, W. J., Said, T. A., & Hakim, O. F. (n.d.). Resource use conflicts between agricultural production and Jozani - Chwaka Bay National Park Authority. Unpublished study by the Department of Commercial Crops, Fruits and Forestry, Zanzibar.

Kamuzinzi, P. K., Shukla, J., & Ndabaga, E. (2015). The effectiveness of Rwanda Development Board tourism revenue sharing program towards local community socio-economic development: A case study of Nyungwe National Park. *European Journal of Hospitality and Tourism Research*, *3*(2), 47–63.

Khatib, A. H. (2000). Ecotourism in Zanzibar, Tanzania. In P. U. C. Dieke (Ed.), *The political economy of tourism development in Africa* (pp. 167–180). Elmsford, NY: Cognizant Communication Corporation.

Kraus, F., Merlin, C., & Job, H. (2014). Biosphere reserves and their contribution to sustainable development. *Zeitschrift für Wirtschaftsgeographie*, *58*(2–3), 164–180. doi:10.1515/zfw.2014.0011

Lama, A. K., & Job, H. (2014). Protected areas and road development: Sustainable development discourses in the Annapurna Conservation Area, Nepal. *Erdkunde*, *68*(4), 229–250. doi:10.3112/erdkunde.2014.04.01

Lapeyre, R. (2011). Governance structures and the distribution of tourism income in Namibian Communal Lands: A new institutional framework. *Tijdschrift Voor Economische En Sociale Geografie*, *102*(3), 302–315. doi:10.1111/j.1467-9663.2011.00665.x

Leung, Y., Spenceley, A., Hvenegaard, G., & Buckley, R. (2018). Tourism and visitor management in protected areas: Guidelines for sustainability. Best Practice Protected Area Guidelines Series No. 27. Gland: IUCN.

MacKenzie, C. A., Salerno, J., Hartter, J., Chapman, C. A., Reyna, R., Tumusiime, D. M., & Drake, M. (2017). Changing perceptions of protected area benefits and problems around Kibale National Park, Uganda. *Journal of Environmental Management*, *200*, 217–228. doi:10.1016/j.jenvman.2017.05.078

Makame, M. K., & Boon, E. K. (2008). Sustainable tourism and benefit-sharing in Zanzibar: The case of Kiwengwa-Pongwe forest reserve. *Journal of Human Ecology*, *24*(2), 93–109. doi:10.1080/09709274.2008.11906105

Manuel-Navarrete, D. (2016). Tourism and sustainability. In H. Heinrichs, P. Martens, G. Michelsen & A. Wiek (Eds.), *Sustainability science* (pp. 283–291). Dordrecht: Springer Netherlands.

Mayer, M., & Job, H. (2014). The economics of protected areas—A European perspective. *Zeitschrift für Wirtschaftsgeographie*, *58*(1), 73–97. doi:10.1515/zfw.2014.0006

Mayring, P. (2015). *Qualitative Inhaltsanalyse: Grundlagen und Techniken* (12th ed.). Weinheim, Basel: Beltz.

Monshausen, A., Tremel, C., Plüss, C., Koschwitz, G., & Lukow, M. (2016). *Transforming Tourism: The 2030 Agenda for sustainable development*. Berlin: Bread for the World—Protestant Development Service. Retrieved from http://www.tourism-watch.de/files/2030_agenda_internet_en_0.pdf

Morais, D. B., Bunn, D., Hoogendoorn, G., & Kc, B. (2018). The potential role of tourism microentrepreneurship in the prevention of rhino poaching. *International Development Planning Review*, *40*(4), 443–461. doi:10.3828/idpr.2018.21

Mukanjari, S., Bednar-Friedl, B., Muchapondwa, E., & Zikhali, P. (2013). Evaluating the prospects of benefit sharing schemes in protecting mountain gorillas in Central Africa. *Natural Resource Modeling*, *26*(4), 455–479. doi:10.1111/nrm.12010

Munanura, I. E., Backman, K. F., Hallo, J. C., & Powell, R. B. (2016). Perceptions of tourism revenue sharing impacts on Volcanoes National Park, Rwanda: A sustainable livelihoods framework. *Journal of Sustainable Tourism*, *24*(12), 1709–1726. doi:10.1080/09669582.2016.1145228

OCGS (2016). *Office of chief government statistician*. Retrieved from http://www.ocgs.go.tz/

Saldaña, J. (2009). *The coding manual for qualitative researchers*. Los Angeles, Sage.

Salerno, F., Viviano, G., Manfredi, E. C., Caroli, P., Thakuri, S., & Tartari, G. (2013). Multiple carrying capacities from a management-oriented perspective to operationalize sustainable tourism in protected areas. *Journal of Environmental Management*, *128*, 116–125.. doi:10.1016/j.jenvman.2013.04.043

Salum, L. A. (2009). Ecotourism and biodiversity conservation in Jozani-Chwaka Bay National Park, Zanzibar. *African Journal of Ecology*, *47*, 166–170. doi:10.1111/j.1365-2028.2008.01065.x

Sandbrook, C., & Adams, W. M. (2012). Accessing the impenetrable: The nature and distribution of tourism benefits at a Ugandan National Park. *Society & Natural Resources*, *25*(9), 915–932. doi:10.1080/08941920.2011.644394

Saner, R., Yiu, L., & Filadoro, M. (2015). Tourism development in least developed countries: Challenges and opportunities. In A. Camillo (Ed.), *Handbook of research on global hospitality and tourism management* (pp. 229–255). Hershey: Information Science Reference.

Schmidt-Soltau, K., & Brockington, D. (2007). Protected areas and resettlement: What scope for voluntary relocation? *World Development*, *35*(12), 2182–2202. doi:10.1016/j.worlddev.2007.02.008

Schnegg, M., & Kiaka, R. D. (2018). Subsidized elephants: Community-based resource governance and environmental (in)justice in Namibia. *Geoforum*, *93*, 105–115. doi:10.1016/j.geoforum.2018.05.010

SMZ (2013). *Tourism in Zanzibar*. Retrieved from http://www.zanzibar.go.tz/index.php?rgo=tourism

Snyman, S. (2017). The role of private sector ecotourism in local socio-economic development in southern Africa. *Journal of Ecotourism*, *16*(3), 247–268. doi:10.1080/14724049.2016.1226318

Spenceley, A., Habyalimana, S., Tusabe, R., & Mariza, D. (2010). Benefits to the poor from gorilla tourism in Rwanda. *Development Southern Africa*, *27*(5), 647–662. doi:10.1080/0376835X.2010.522828

Spenceley, A., Snyman, S., & Rylance, A. (2017). Revenue sharing from tourism in terrestrial African protected areas. *Journal of Sustainable Tourism*, 1–15. doi:10.1080/09669582.2017.1401632

Swemmer, L., Grant, R., Annecke, W., & Freitag-Ronaldson, S. (2015). Toward more effective benefit sharing in South African National Parks. *Society & Natural Resources*, *28*(1), 4–20. doi:10.1080/08941920.2014.945055

Tapper, R. (2007). *Managing tourism & biodiversity: User's manual on the CBD guidelines on biodiversity and tourism development*. Montreal: CBD.

TIES (2015). *What is ecotourism?* Retrieved from https://www.ecotourism.org/what-is-ecotourism

Tumusiime, D. M., & Vedeld, P. (2015). Can biodiversity conservation benefit local people? Costs and benefits at a strict protected area in Uganda. *Journal of Sustainable Forestry*, *34*(8), 761–786. doi:10.1080/10549811.2015.1038395

Tumusiime, D. M., & Vedeld, P. (2012). False promise or false premise? Using tourism revenue sharing to promote conservation and poverty reduction in Uganda. *Conservation and Society*, *10*(1), 15–28. doi:10.4103/0972-4923.92189

UN. (2015). *Transforming our world: The 2030 agenda for sustainable development (A/RES/70/1)*. New York: Author. Retrieved from https://sustainabledevelopment.un.org/content/documents/21252030%20Agenda%20for%20Sustainable%20Development%20web.pdf

UNEP & UNWTO. (2005). *Making tourism more sustainable: A guide for policy makers*. Paris: Author.

UNESCO MAB ICC. (2016). 28th Session, Lima/Peru, 18–19 March, Final Report (No. SC-16/CONF.228/13). Paris: UNESCO. Retrieved from http://www.unesco.org/new/fileadmin/MULTIMEDIA/HQ/SC/pdf/SC-16-CONF-228-12_final_en_v2.pdf

UNESCO WHC. (1997). *Jozani—Chwaka Bay conservation area.* Retrieved from http://whc.unesco.org/en/tentativelists/850/

UNWTO (2013). *Sustainable tourism for development guidebook: Enhancing capacities for sustainable tourism for development in developing countries.* Madrid: Author.

UNWTO (2015). *Tourism and the sustainable development goals.* Madrid: Author. Retrieved from http://cf.cdn.unwto.org/sites/all/files/pdf/sustainable_development_goals_brochure.pdf

USAID (1996). *Conducting key informant interviews.* Washington, DC: USAID Center for Development Information and Evaluation. Retrieved from http://pdf.usaid.gov/pdf_docs/pnabs541.pdf

Index

Page numbers in **bold** refer to tables and those in *italic* refer to figures.